Multid[illegible] Approach and Safe[illegible]oled by Buoyancy Driven Flows

Multidisciplinary Design Approach and Safety Analysis of ADSR Cooled by Buoyancy Driven Flows

Proefschrift

ter verkrijging van de graad van doctor
aan de Technische Universiteit Delft
op gezag van de Rector Magnificus prof.dr.ir. J. T. Fokkema
voorzitter van het College voor Promoties,
in het openbaar te verdedigen op maandag 5 februari 2007 om 12:30 uur

door

Carlos Alberto CEBALLOS CASTILLO

Werktuigkundig ingenieur
Technische Universiteit Delft
geboren te Bogota DC, Colombia

Dit proefschrift is goedgekeurd door de promotoren:
Prof. dr.ir. A.H.M. Verkooijen
Prof. dr.ir. T.H.J.J. Van der Hagen

Samenstelling promotiecommissie:

Rector Magnificus	Voorzitter
Prof.dr.ir. A.H.M. Verkooijen	Technische Universiteit Delft, promotor
Prof.dr.ir. T.H.J.J. Van der Hagen	Technische Universiteit Delft, promotor
Prof.dr.ir. R.F. Mudde	Technische Universiteit Delft
Prof.dr.ir. I.M. Richardson	Technische Universiteit Delft
Prof.dr. J. Martinez-Val	ETSII, Madrid, Spanje
Dr. H. Ait Abderrahim	Studiecentrum voor Kernenergie, Mol, Belgie
Dr.ir. D. Lathouwers	Technische Universiteit Delft

ISBN 978-1-58603-720-8

Keywords: ADS, ADSR, Natural Circulation, Liquid Metal, Buoyancy, Inherent safety, Safety analysis, Transients, Subcritical reactor, Gas Lift, Creep-fatigue, Transmutation

Published and distributed by IOS Press under the imprint Delft University Press

Publisher
IOS Press
Nieuwe Hemweg 6b
1013 BG Amsterdam
The Netherlands
Tel: +31-20-688 3355
Fax: +31-20-687 0019
Email: info@iospress.nl
www.iospress.nl
www.dupress.nl

PRINTED IN THE NETHERLANDS

Dedicated to my beloved parents

Summary

Nuclear waste is composed of transuranic elements (plutonium and minor actinides) and fission products. The separation of plutonium from minor actinides in the nuclear waste offers a more efficient utilization of resources because plutonium can be recycled as fuel in different reactors. Similarly, the transformation of the minor actinides in lighter elements with transmutation reactions is a method to reduce the storing time and the amount of waste to be stored in the geological repository.

Transmutation can be achieved in all types of reactors - thermal systems, fast systems, critical and sub-critical systems (Gudowski et al., 2001). Fast spectrum systems have significant advantages because they offer higher transmutation efficiency in comparison to thermal systems (OECD/NEA, 2002). However, the addition of actinides to the fuel has adverse effects on safety parameters: the fraction of delayed neutrons and the Doppler coefficient are reduced (Eriksson et al., 2005). This parameter is very important to assure the dynamic control and safe operation of a critical reactor and for this reason the actinide content in the fuel of critical reactors has to be limited.

Alternatively, if the reactor is subcritical and an external neutron source is supplied, the system would be able to operate with a steady power level without a self-sustained chain reaction (Rubbia et al., 1995). In this way, the shutdown of the external source gives the possibility for rapid control of any undesired power excursion in the transmuter. This novel reactor concept has been introduced as the Accelerator Driven Subcritical Reactor (ADSR).

The ADSR operating principle consists of feeding a subcritical core with an external neutron source. The neutron source is obtained from powerful collisions of a high-energy charged particles with a high Z target. The high-energy particles are protons produced by a particle accelerator, the protons travel through a beam tube and collide with a target. The spallated neutrons will start a chain reaction in the nuclear core and the rate will be dependent on the proton flux from he accelerator. The spallation target would be located in the middle of the reactor core. In this way, effective control can be executed through the accelerator's proton current. The use of ADSR in the fuel cycle offers interesting degrees of freedom in the core design, which can maximize the effectiveness of transmutation. Another advantage is that almost any fuel composition including those with low safety characteristics can be tolerated. ADSR is a promising technology that may benefit the society from different viewpoints. However, there are many technical issues to be solved before an ADSR system is realized.

The development of the ADSR depends on the successful integration of different systems, for which a safety strategy has to be adapted. This dissertation contributes to the development of the ADSR technology by studying two important safety aspects: the time-dependent response during transient conditions and its influence over the structural integrity. For this purpose, thermalhydraulics, neutronics and structural models of the reactor are integrated.

The results have shown that a subcritical reactor is remarkably effective to overcome transients involving criticality risk. However, core damage and/or melt down are evident if the beam is not shutdown. Negative reactivity feedbacks do not reduce the power significantly and since, large percentage of heat is not efficiently removed, the structural integrity is exposed.

Natural circulation is not helpful in view of the fact that the mass flow rate is dependent on the density difference between the hot and cold legs. During transients, it can occur than the hot leg becomes cooler and the cold leg gets hotter (e.g. during a a loss of heat sink transient), in this case, the buoyancy force changes direction working against the inertial force. If natural circulation is enhanced with a gas-lift pump, the power rate of the reactor can be increased. Still, the oscillatory nature of the loop system is pronounced by gas density changes during temperature transients. This, subjects the mechanical components to higher thermal stress cyclic rates increasing the fatigue damage.

Another technical disadvantage of the buoyancy driven liquid metal reactors becomes the large temperature difference across the core, which increase the severity of thermal shocks. This is especially critical when looking at the beam trip frequency of current accelerators. A non-conservative estimate shows that the reliability of the accelerator should be increased by a factor of 5 in order to avoid cladding failure during the first year of operation.

In conclusion, the inherent safety attached to the subcriticality in an ADSR presents large benefits to avoid criticality accidents. Nevertheless, the development challenges for this technology remain in the core power control and the accelerator reliability. Regarding natural circulation and enhancement with gas lift pump, the maximum core power is limited and the transient behavior does not satisfy the safety requirements for all scenarios.

Samenvatting

Kernafval bestaat uit transuranen (plutonium en "minor actinides", hierna aangeduid als MA) en splijtingsproducten. Het scheiden van plutonium van MA in het kernafval leidt tot een efficiënter gebruik van grondstoffen omdat plutonium gerecycleerd kan worden in diverse/verschillende (typen) reactoren. Tevens is de transmutatie van MA tot lichtere elementen een methode om de benodigde opslagtijd en de hoeveelheid afval binnen een geologische opslag te verkleinen.

Transmutatie kan worden bereikt in diverse reactortypen – thermische systemen, snelle systemen, kritieke en subkritieke systemen (Gudowski *et al.*, 2001). Snel-spectrum systemen hebben grote voordelen omdat deze een groter transmutatievermogen hebben in vergelijking met thermische systemen (OECD/NEA, 2002). Echter, de toevoeging van actiniden aan de splijtstof heeft een negatief effect op veiligheidsgerelateerde parameters: de fractie vertraagde neutronen en de Doppler coëfficiënt zijn kleiner (Eriksson *et al.*, 2005). Deze parameters zijn bijzonder belangrijk om de dynamische controle en veilig bedrijf van een kritieke reactor en vanwege dit feit dient de actinide concentratie in de splijtstof beperkt te blijven. In geval van een subkritiek systeem waar een externe neutronenbron wordt gebruikt kan het systeem een constant vermogen leveren zonder een zichzelf in stand houdende kettingreactie (Rubbia *et al.*, 1995). Afschakeling van de externe neutronenbron kan dan op korte termijn eventueel optredende vermogensoscillaties controleren. Dit reactor concept is eerder geïntroduceerd onder de naam "Accelerator Driven Subcritical Reactor" (ADSR).

Het principe van de ADSR bestaat uit een subkritieke kern die gevoed wordt door een externe neutronenbron. Deze bron wordt verkregen door hoogenergetisch geladen deeltjes te schieten op een trefplaat gemaakt van materiaal met een hoge Z (atoomgetal). De hoogenergetische deeltjes zijn protonen uit een deeltjesversneller. De zogenaamde spallatieneutronen geven een cascade van neutronen waarbij de hoeveelheid afhangt van de protonflux van de versneller. De spallatie trefplaat bevindt zich in het centrum van de subkritieke kern; dit leidt tot de grootste neutronenflux door minimalisatie van lek. Het gebruik van de ADSR in de brandstofcyclus geeft vrijheid in het ontwerpen van een reactorkern welke het transmutatievermogen kan maximaliseren. Een ander voordeel is dat vrijwel iedere brandstofcompositie getolereerd kan worden, inclusief die met minder dan optimale veiligheidskarakteristieken.

In dit proefschrift wordt een bijdrage geleverd aan de ontwikkeling van ADSR technologie door twee veiligheidsaspecten te bestuderen: het tijdafhankelijk gedrag onder diverse condities en de invloed hiervan op de structurele integriteit. Hiervoor zijn modellen voor neutronica, thermohydraulica en mechanische analyse geïntegreerd.

De resultaten tonen aan dat een subkritieke reactor opmerkelijk effectief is in het ondervangen van ongevalscenario's met een criticiteitsrisco. Schade aan de reactorkern of het smelten ervan kan echter voorkomen als de bron niet tijdig afgeschakeld wordt. Negatieve terugkoppeling van de reactiviteit reduceert het vermogen niet significant en omdat een groot gedeelte van de warmte niet afgevoerd wordt de mechanische stabiliteit negatief beïnvloed.

Natuurlijke circulatie is geen uiteindelijke oplossing vanwege het feit dat de massastroom afhankelijk is van de hete en koude secties. Gedurende transiënten kan het voorkomen dat de hete sectie kouder wordt dan de koude sectie (bv gedurende een ongeval met verlies van warmteafvoer) waardoor de drijvende kracht van richting veranderd en de massastroom doet afnemen. Wanneer natuurlijke circulatie wordt verhoogd met een "gas-lift" pomp is het mogelijk het reactorvermogen te verhogen. Het oscillerende gedrag van de koellus is echter sterk aanwezig gedurende temperatuur transiënten vanwege dichtheidveranderingen van het ingebrachte gas. Deze temperatuursveranderingen stellen de mechanische componenten bloot aan thermische stress waardoor de kans schade door moeheid toeneemt.

Een ander technisch nadeel van een natuurlijke circulatie vloeibaar metaal gekoelde reactor is het grote temperatuurverschil over de kern waardoor de hevigheid van thermische schokken toeneemt. Dit is met name belangrijk i.v.m. de frequentie van versnellerafschakelingen. Een voorzichtige schatting leert dat de betrouwbaarheid van versnellers met een factor 5 dient toe te nemen om falen van de cladding te voorkomen gedurende het eerste jaar van reactorbedrijf.

Concluderend leidt de inherente veiligheid verbonden aan de subcriticiteit van de ADSR tot grote voordelen bij het voorkomen van criticiteitsongevallen. Er blijven echter grote uitdagingen op het gebied van controle van het kernvermogen en de betrouwbaarheid van de versneller. Op het gebied van natuurlijke circulatie en de verbetering hiervan m.b.v. de "gas-lift" pomp kan geconcludeerd worden dat het kernvermogen gelimiteerd is en dat de veiligheidseisen niet onder alle omstandigheden gegarandeerd kunnen worden.

Contents

"The release of atomic energy has not created a new problem, it has merely made more urgent the necessity of solving an existing one"

Albert Einstein, 1879 -1955

1 Introduction

The energy consumption in the world is directly related to the economical growth. The International Energy Agency estimates a doubling increase of electricity demands in the next 25 years (Energy Outlook, 2004). Somehow, this growth must be satisfied without destroying the ecological equilibrium of our planet. For this reason, all sorts of energy resources, which are potential pollutants for the environment are being investigated and their technologies improved, as well as new renewable energies are developed. It is also evident that renewables will take several decades before they are able to substitute the present capacity and even more, to cope with the future energy demands.

1.1 Sustainable Nuclear Energy

Nuclear energy is an air pollution free technology with the potential to satisfy the world's energy demands for many centuries. However, some concerns about the use of nuclear energy have to be further developed in order to recognize nuclear energy as a sustainable option. These concerns are the use of nuclear energy for weapons, the risk of accidents with radioactivity release and the waste management.

Nuclear waste is composed of trans-uranic (TRU) elements and fission products. The TRU are the result of neutron capture in the fuel and subsequent decay. They can be separated in Pu and minor actinides (MA). The fission products are fragments produced during the fission process. Most of the fission products are short-lived (less than 300 years) in comparison with the TRU elements, which can take thousands of years to reach the reference radiotoxicity level (ENEA, 2001).

Two methods are considered for managing radioactive waste: *i)* geological disposal, which consists of isolating the waste from the biosphere by placing it in a safe geological formation and wait for it to decay. The geological formation and a good container design have to assure the isolation of waste for a long period of time. These issues raise uncertainty about the future circumstances and the economical sustainability of this solution. *ii)* the separation of plutonium from MA and transformation of the MA in lighter elements with transmutation reactions before storing. The lighter elements have shorter half-lives and in this way, they would need to stay for a considerably shorter time in the repository. At the same time, the generation of electricity from MA is a more efficient utilization of resources.

All types of reactors can be used to transmute the MA - thermal systems, fast systems, critical and sub-critical systems (Gudowski *et al.*, 2001). Fast spectrum systems have significant advantages because they offer higher transmutation efficiency in comparison to thermal systems (OECD/NEA, 2002). However, the addition of actinides to the fuel has adverse effects on safety parameters: the fraction of delayed neutrons and the Doppler coefficient are reduced (Eriksson *et al.*, 2005). These parameters are very important to assure the dynamic control and safe operation of a critical reactor and for this reason, the actinide content in the fuel of critical reactors has to be limited.

Alternatively, if the reactor is subcritical and an external neutron source is supplied, the system would be able to operate with a steady power level without a self-sustained chain reaction (Rubbia *et al.*, 1995). In this way, the shutdown of the external source gives the possibility for rapid control of any undesired power excursion in the transmuter. Last, the possibility of Plutonium recovery within a closed and controlled fuel cycle minimizes the risk of using it for weapons proliferation.

1.2 ADSR Role in the Nuclear Fuel Cycle

The transmutation efficiency is very important in any fuel cycle strategy. The efficiency is connected to the energy spectrum, burn-up time and cost to transmute the waste. Fast reactors and ADSR will play an important role in this regard. The selected fuel cycle strategy may be suited to different performance requirements like the flexibility, the technological requirements, the proliferation, the economics, etc. A study carried out by OECD/NEA (2002) compared two waste management approaches: one considers plutonium as a by-product from light water reactors (LWR) and separates it from the minor actinides stream. The other approach re-processes them together. The second option is attractive because it enhances proliferation resistance but the first option offers a more economic competitiveness. These two approaches can be accommodated in different fuel cycle strategies according to technological and/or economical challenges. The schemes investigated by OECD/NEA are presented in figure 1.1, the figure shows 5 different transmutation strategies relevant to the future requirements.

The first option is to remain on the once-through fuel cycle. This is a good option for countries with a modest nuclear energy program, since the amount of waste does not compensate the investments of reprocessing. A second strategy is to close the fuel cycle for Pu by recycling first in LWR and then in fast reactors (FR). Since the U and Pu are separated, the radiotoxicity level of waste is reduced considerably, but the actinide stream will go to the high level waste (HLW) repository. A third option considers transmutation to close the fuel cycle directly for LWR using either an ADSR or a FR. In this way, the actinides recycling will highly reduce the radiotoxicity content of radioactive waste. A fourth option has been proposed to overcome the adverse effects of actinides in the fuel. This is the so called "Double Strata" strategy. In the first stratum, energy is produced using conventional reactors. Pu may or may not be recycled in this stratum. "Dedicated" reactors will be used in the second stratum to burn actinides and some Pu, which no longer can be used in the first stratum. The last option is a fast reactor strategy substituting the LWR park. It will offer a closed Pu fuel cycle with actinide recycling, however, the elimination of the LWR it is not easily projected in short term view.

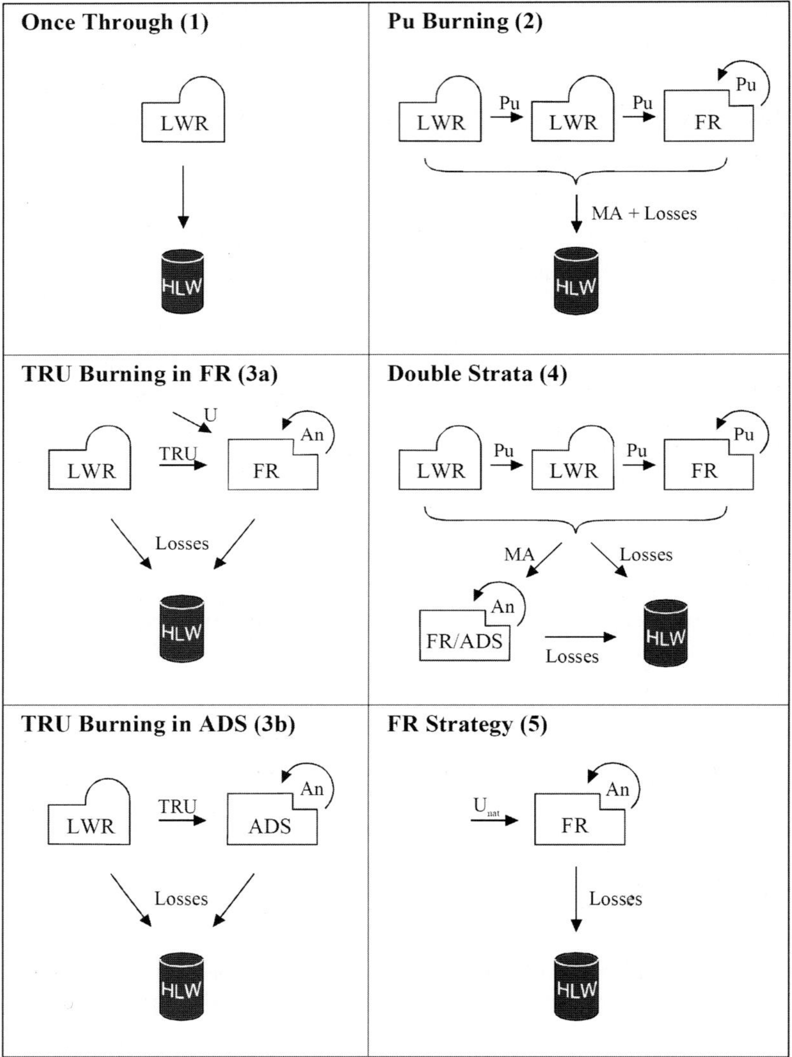

Figure 1.1 Fuel cycle schemes considered in the study by NEA (source: OECD/NEA, 2002) LWR: light water reactors, FR: fast reactors, MA: minor actinides, HLW: high level waste, ADS: accelerator driven systems, U: Uranium, Pu: Plutonium

Important conclusions of NEA's investigation are that all transmutation strategies with multiple recycling of the fuel can achieve similar radiotoxicity reduction in the long-term waste radiotoxicity. These strategies can achieve a hundred reduction factor in the long-term waste radiotoxicity and even higher for actinides inventory reduction. Partially closed cycles are a near-term transmutation option, but they do present a factor of two less reduction than the fully closed strategies. Fully closed cycles might be realised with a 10-20% increase of electricity cost. For partially closed cycles the increase is about 7% of the electricity cost with regards to the once through cycle.

In a similar study, Cometto *et al.*, (2004) have presented comparable results. They suggest that due to its economical and technological requirements, ADSR's are best suited to work as minor actinide burners in the double strata cycle, whereas critical FR's are better for Pu and MA together in a TRU burning strategy. Another similar study by Hoffman and Stacey, (2002) declares that an ADSR would be capable of a net TRU destruction rate 2 to 3 times larger than a critical FR (similar to what NEA has reported). This advantage is only visible in the use of non-fertile TRU in the ADSR. On the other hand, if energy is produced, the ADSR would account for 25% of the total generated power while the FR would produce a 45% of the total power.

The use of ADSR in the fuel cycle offers interesting degrees of freedom in the core design, which can maximise the effectiveness of transmutation. Another advantage is that almost any fuel composition including those with low safety characteristics can be tolerated. Other important developments that ADSR technology brings to the society can be electricity production, source of isotopes for medical, industrial and research purposes, potential to generate significantly higher fluxes for material research. ADSR is a promising technology that may benefit the society from different viewpoints. However, there are many technical issues to be solved before an ADSR system is realised. Active R&D is needed to bring this option under way.

1.3 General Description & Key Developments

This novel reactor concept was proposed by Rubbia *et al.*, (1995). It consists of feeding a subcritical core with an external neutron source. The neutron source is obtained from powerful collisions of a high-energy charged particles with a high Z target. The high-energy particles are protons produced by a particle accelerator, the protons travel through a beam tube and collide with a target.

The spallated neutrons will start a chain reaction in the nuclear core and the rate will be dependent on the proton flux from the accelerator. The spallation target would be located in the middle of the reactor core. In this way, effective control can be executed through the accelerator's proton current.

A schematic drawing of an ADSR system is shown in figure 1.2, identifying its major components. A high power particle accelerator produces energetic protons that hit a heavy metal target to produce neutrons. The neutrons feed the nuclear fission process of a subcritical reactor where nuclear waste has been placed. A conventional power cycle is added to remove heat and generate electricity.

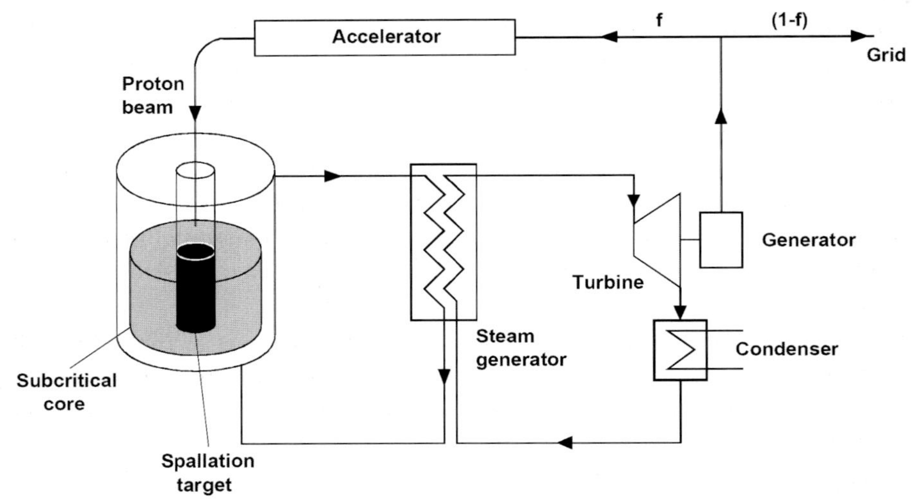

Figure 1.2 Scheme of the ADSR (source: OECD/NEA,2002)

Different ideas have been proposed for ADSR concepts based on fundamental physical properties such as neutron energy spectrum - fast and thermal (Gudowski *et al.*, 2001), fuel type - solid, liquid (Takano and Nishara, 2002; Nifenecker *et al.*, 2003) and coolant – lead-bismuth eutectic (LBE), lead, sodium, and gas (Kerkdraon *et al.*, 2003; Nifenecker *et al.*, 2001). Most international programs seem to be evolving toward fast-spectrum, liquid metal-cooled sub-critical assemblies driven by large linear accelerators (Sasa *et al.*, 2004; ENEA, 2001). For target materials, heavy metal such as lead-bismuth (Pb-Bi) or tungsten are proposed (Seltborg *et al.*, 2003; Vickers, 2003).

Although accelerator development has advanced and linear accelerators are capable to accelerate protons to several GeV, the frequently repeated beam trips can significantly damage the reactor structures, the spallation target and the fuel, decreasing the ADSR plant availability. The other reliable option for high-power beams is to use cyclotrons, however, they are limited to maximum energies of about 1 GeV and electric currents of a few mA. This may be sufficient to drive a small ADSR but insufficient for large applications (Ponomarev, 2002). The accelerator also has to be capable to produce a stable and reliable low and high intensity proton beam, which is needed for the start process and the steady operation. The reliability requirements are related to the number of allowable beam trips and its effect on plant parameters deviation (thermal power, primary flow, pressure, temperature). The spallation target design for an ADSR should optimize the neutronic efficiency, the material properties and thermal-hydraulics, since it is simultaneously subject to severe thermal-mechanical loads and damage due to high-energy heavy particles - protons, spallation and fission neutrons (Tak *et al.*, 2005).

The selected subcriticality level must be properly balanced considering economical reasons like the desire to construct a system with a low beam power, and safety reasons like that a small subcriticality level implies an increased risk of approaching criticality under transient conditions, but also larger effects of the negative feedbacks. An adequate level of subcriticality can be achieved by conservatively estimating the positive reactivity insertions. They are associated with incidents and accident conditions such as fuel, coolant and structural materials temperatures variation, coolant voiding and properly choosing the allowable range of normal operating conditions (Perdu *et al.*, 2003). An additional requirement is the compensation of fissile material burn-up indicating that an operational range to maintain the steady required power level is needed, as well as to execute a careful reactivity monitoring to assure safe operation (Baeten and Abderrahim, 2004; Bianchi *et al.*, 2005).

The fast neutron spectrum can be obtained by selecting liquid metals for primary coolant like (Pb, Pb-Bi, Na) or gases (He, CO_2). Liquid metals are good candidates due to their good thermal-physical properties and the possibility to operate at atmospheric pressure (Morita *et al.*, 2004). On the other hand, they exhibit disadvantages such as corrosion for Pb and Pb-Bi or strong chemical reactivity with air and water for Na, as well as possible positive reactivity feedback from voiding. They also cause difficulties with regard to in-service inspection and repair due to the opacity (Abderrahim *et al.*, 2002). Their melting points are disadvantageous during shutdown and refuelling since primary coolant freezing must be avoided.

The core can be cooled either by forced or natural circulation. The use of natural circulation is preferred since it eliminates the occurrence of loss of flow accidents and the cost involved with the operation and maintenance of redundant systems. Natural circulation was an important inherent safety feature proposed in the first concept of ADSR (Rubbia *et al.*, 1995) and it is being considered to implement it in the first experimental XADS facility (Ansaldo, 2001; Cinotti, 2004). The ADSR can be connected to a conventional power cycle with an intermediate cooling circuit or directly to the steam generator. The design of the intermediate heat exchanger or steam generator and placement inside the reactor vessel is constraint by the vessel size, which is equally constraint by the capital cost and the compactness desired features of such a plant (Salvatores, 2005).

Regarding fuels: oxides and nitrides are considered as the most promising fuel materials (ENEA, 2001). Oxide phases have the advantage of high chemical stability and relative simple handling and fabrication, which is very important for actinides handling. However, the relatively low thermal conductivity of oxide materials leads to high operating temperatures. Composites (ceramic-metal or ceramic-ceramic) are preferred to provide the high heat transfer rates required to avoid large peaking temperature in the fuel: composite with steel (CERMET) or MgO (CERCER) are the first referenced options for advanced fuels (Eriksson *et al.*, 2005; Chen *et al.*, 2004; Maschek *et al.*, 2003). Using liquid fuels would avoid burn-up reactivity changes by adding fissile material on-line and removing poisons. However, compared to solid fuels, liquid fuels present much larger unknowns associated with material compatibility and operation. R&D on these key issues and the system integration are the most important steps towards the ADSR demonstration. These developments have to be realized with economic competitive advantages and maximizing safety characteristics like inherent and passive safety systems.

1.4 Motivation

The development of an ADSR depends extensively on the successful integration of different systems initially conceived for a different purpose such as the particle accelerator, the spallation source and the reactor core. During the design and development process, a safety philosophy has to be pursued. The basic requirements for this philosophy have to cope with the general guidelines of nuclear reactor safety – the protection of people and the environment by establishing and maintaining effective defenses against radiological hazards (IAEA, 1993).

The best procedure for optimizing the design is a conscientious safety analysis of the system. The most traditional technique for verifying and demonstrating the safety of any nuclear facility is the defence in depth (IAEA, 1996). It is implemented by listing the initiating events (internal and external hazards) and classify them by categories regarding their expected frequency of occurrence. The events are analyzed with stringent rules. The prevention has to be pursued by passive control, active actions and inspection activities. As a second step, the mitigation of the accident is required despite the high prevention levels are achieved by the first analysis. The process is complemented by a line of defence method which defines the safety requirements for the safety systems. They can be divided on three types: the preventive measures to avoid initiating events, the active and passive measures and the inherent behavior or resistance by natural behavior.

Recent strategies to reduce or exclude potential accidents and improve reactor economy involve the use of buoyancy driven flows in the reactor primary cooling circuit. In principle, the implementation of natural circulation in a liquid metal cooled nuclear reactor could suppress the dependence on external pumps and assure a safe and reliable operation. Even in a forced circulation system, the capability of the coolant to develop natural circulation is of great interest for decay heat removal of the core after reactor shutdown. The advanced LWR adopts natural circulation as its cooling principle and are projected to cover the small and medium power ranges from the LWR park. Liquid metal fast reactors (LMFR) and ADSR cooled by natural circulation seem to have a more narrow range of operation (Davis *et al.*, 2002; Chang *et al.*, 2000), as the coolant does not undergo phase change in the core. Therefore, the temperature difference across the core of a liquid metal reactor exceeds by far the temperature difference of the LWR and its technical demands to the structural components become more stringent.

Previous work on the assessment of natural circulation of liquid metals in a reactor pool was performed by Rubbia *et al.*, (1995) during the design phase of the ADSR. They have used one-dimensional modeling to simulate the flow and point kinetics theory with a feedback kernel model to consider the ADSR neutron kinetics. The results were optimistic and the idea is being further developed as the experimental XADS facility from the European Union (Ansaldo, 2001). Analysis of transients has also been carried out for the XADS (Coddington *et al.*, 2004). Simplified models for the two-phase flow in the riser and detailed modeling of the reactor vessel auxiliary cooling system (RVACS) have shown that the delay before beam shut-off after initiation of loss of heat sink accident is a critical parameter (Carlsson and Wider, 2002). The successful utilization of natural circulation in a large scale ADSR will imply the reduction of risks during operation. Hence, it is of great interest to study the ADSR design and safe operation focusing on the possible implementation of a natural circulation system for an industrial scale ADSR.

Regarding the dynamics of a sub-critical system, the first concern is the control of power and monitoring of reactivity. The system needs to shut down the beam if an unwanted variation of core parameters is detected (neutron flux, primary flow, temperature). The reactor has to remain subcritical at any conceivable state, an adequate margin to criticality is needed for any event like the positive reactivity insertions associated with changing conditions from fuel, coolant and structural materials temperatures variations. In the event of unprotected transients (beam remains on) a reliable safety related beam shutdown system is desirable (Eriksson and Cahalan, 2002), as well as a long grace time period for the system to act. Transient analysis of the ADSR would allow us to understand and predict the reactor response under different scenarios threatening the ADSR safe operation.

The other safety concern relates to the assessment of the ADSR structural integrity. The safety level is achieved by implementing an optimal design with good performance for operating at nominal power during its lifetime. In case of power generation, it also has to be capable to operate for long periods at partial load without major impact on safety and performance. The present nuclear plants accept 1 or 2 unexpected shutdowns per year. Existing accelerators exceed these numbers by far. This means that operating an accelerator at a high beam power and requiring few beam trips is a technological challenge. For an industrial ADSR, the tolerable number of beam interruptions has to be reduced. On the other hand, the effect of short and long beam trips has to be recognized and the maximum limit of trips for controlling the thermal-mechanical loading on the fuel assemblies and other components identified. This information is valuable for the improvement of the accelerators reliability and availability.

1.5 Objectives and Scope

The goal is to contribute to the development of ADSR technology by studying two main safety aspects: the transient response and its impact on the structural integrity. For this purpose, the neutronic, thermalhydraulic and mechanical systems have to be integrated.

The thermalhydraulic system plays an important role in the performance and the safety. Hence, it is a primary objective to develop a better understanding of the natural circulation thermalhydraulic systems for a liquid metal cooled ADSR. The feasibility and reliability of a high power liquid metal cooled buoyancy driven reactor has to be investigated. Likewise, the effects of the thermal coupling with neutron kinetics verified. Additionally, the interaction of the beam with the thermalhydraulic and neutronic systems, as well as with the mechanical components has to be reviewed. This work will develop a better understanding of the ADSR design parameters, characteristic dynamic response and provide information for future optimization and design guidelines.

The methodology employed set the following steps: first, to build a thermalhydraulic model of the ADSR and to couple it with a neutronic dynamic model. Then, to study the thermalhydraulic behavior during steady state and transient conditions, evaluating the modeling results within the framework of safety. Subsequently, to integrate the mechanical systems in the model and assess their structural integrity. As a result, to identify routes for the development of the ADSR technologies and safety guidelines.

1.6 Outline

This thesis presents the work-study on safety analysis of ADSR cooled by buoyancy driven flows. The thesis is outlined as follows:

In chapter 2, the ADSR safety issues are discussed by the relevant physical systems – neutronic, thermalhydraulic and mechanical. From these issues, the safety parameters constraining the reactor design and operation are identified. Then, some margin or limits to these parameters are given and a number of hypothetical scenarios that might threat the ADSR safe operation are selected.

In chapter 3, a detailed description of the systems interaction is described. Physical models for neutronic, fuel pin heat transfer and thermalhydraulic models are presented. The modeling approach and the model validation are discussed as well. Special interest is shown in the thermalhydraulic modeling approaches in which a simplified and a detailed geometrical description of the core have provided different levels of problem understanding and information.

Chapter 4 explores the feasibility of cooling a liquid metal ADSR using buoyancy driven flows. The first part explains the nature of the natural circulation single-phase reactor and its scaling principles. Then, it discusses the implementation of the buoyancy enhancement with gas injection in order to increase the core power capacity. The study takes as a baseline the design from MIT and INL (Davis *et al.*, 2002) for a single-phase natural circulation lead-bismuth actinide burner. At the end of the chapter, some observations are introduced regarding the hydrodynamic effects on flow and cooling performance.

Chapter 5 presents the transient analysis of the scenarios established in chapter 2. The event consequences are determined and valued with the safety criteria that were discussed in chapter 2 as well. The results obtained with a one-dimensional thermalhydraulic model compare two concepts: the single-phase and the two-phase buoyancy driven flow reactors. The results have characterized the dynamic response from every system. Then, a detailed two-dimensional model of the core demonstrates the safety advantages of the two-phase concept and provides information for the structural integrity assessment of the fuel pin cladding.

Chapter 6 evaluates the impact of beam trips frequency on the ADSR lifetime. The first section describes a thermal-structural stress model suggested to study creep-fatigue interaction in the fuel pin cladding. Then, thermal-stress transient analysis is carried out together with creep analysis and the results are discussed. The structural integrity assessment is derived with engineering rules for reactor design recommended by the ASME Boiler and Pressure Vessel Code – Case N47.

Finally, chapter 7 gives some conclusions and recommendations.

"If I could have just 1% of the money spent on global armaments, no one in this world would go to bed hungry" *James Morris, head of the WFP*

2 Safety Considerations for Design and Control

The development of the ADSR is one of the cases, in which a new technology is introduced by integrating different components conceived with a different purpose. This process requires a systematic approach in order to produce a reliable and safe device. The present chapter introduces the safety related problems of the ADSR, the safety constraints are identified and the safety limits are established as a work frame for the safety assessment. A set of hypothetical scenarios is proposed to support a reactor safety strategy.

2.1 Introduction

The operation principle of the ADSR has been described in detail in section 1.3. By conceptual definition, the ADSR is an inherent safe reactor, because the subcritical state provides a sufficient margin to cope with the reduction on the delayed neutron fraction and the Doppler effect. In case of a positive reactivity insertion, the subcritical state assures that the reactor does not become prompt supercritical and if the controller foresees any risk, it can order the proton beam to stop, bringing the reactor to decay power conditions.

The safety characteristics of the ADSR should also consider different aspects involved in the system integration. For example, the heat removal by the thermalhydraulic system has an effect on neutronic performance during transient conditions. In similar way, the rate of heat removal by the thermalhydraulic system has an effect on structural components lifetime. In addition, the design has to exclude any potential hazards from reactivity excursions and tolerate much more transients than the actual nuclear plants. This is due to the larger amount of beam interruptions.

For study purposes, some relevant parameters were identified, constrained and grouped to form the design limits and safety limits, which describe the safety envelope for the ADSR. These parameters have been grouped according to their nature in: neutronic, thermalhydraulic and mechanical systems constraints.

Figure 2.1 presents the relation between the physical constraints and its entailed effect on reactor safety. The constraints are related to neutron kinetics, thermalhydraulic system and mechanical components. In the present work, the reactor safety is considered from two perspectives: controllability and integrity. The controllability relates to the ability to manipulate safely the neutron multiplication. In critical reactors, the control is achieved by inserting absorbers, permanent monitoring, diversity of actuators and passive measures from design, such as negative feedbacks. In the ADSR, the same function has to be achieved when power evolutions or flow-temperature related problems are detected. The reactor shutdown needs to be achieved and cannot be easily reached if the accelerator is not shutdown. Therefore, the shutdown of the accelerator has to be very reliable and redundant systems, each having an independent function will make a strong line of defence approach. Additionally, controllability should always remain during transients events, normal operating conditions like start up, shutdown, loading and abnormal reactivity evolutions due to burn up. For these purposes, detection and control systems and methods are investigated.

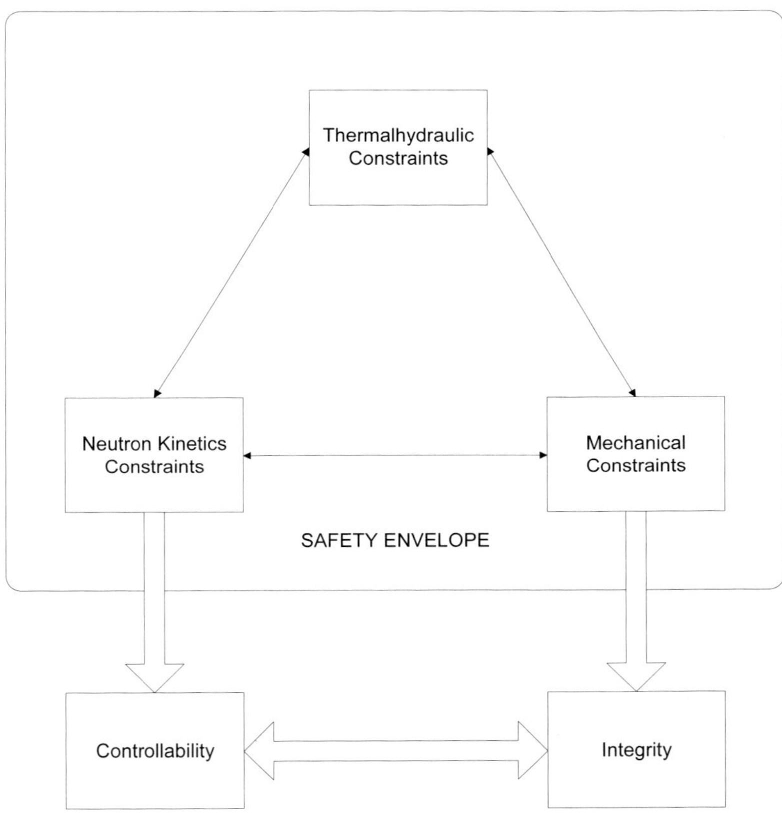

Figure 2.1 Definition of physical constraints and safety envelope for the ADSR operation

The integrity relates to the capacity to withstand any circumstances risking the lifetime and availability of the reactor. The structural components should resist to mechanical and thermal loads and their time-dependent effect during operation and transient events. The loading type could provoke unfavorable or excessive requirements delivering partial or total failure of the component. Moreover, a loss of controllability sometimes has as a consequence on loss of integrity and viceversa. A change in some thermalhydraulic parameters could indirectly lead to controllability or integrity losses, by interaction with the neutron kinetics and mechanical systems.

The safety strategy for the ADSR should provide different mechanisms to influence the constrained parameters (top) and keep them below certain limits. These safety limits define the safety envelope for the safe operation. When the safety limits are exceeded, the safety envelope is abandoned and the controllability and integrity are exposed (bottom).

In the following sections, we discuss the ADSR safety issues using the systems classification from figure 2.1. From these, one should be able to identify the parameters that constraint the reactor safe operation and to differentiate which are design constraints and which, safety constraints. Afterwards, one can introduce margins or limits to these parameters and set a number of hypothetical scenarios for the safety analysis.

2.2 Neutronic Issues

Neutron Kinetics

The primary goal on the design of the ADSR is to maximize the transmutation rate. For this reason, fuel free of ^{238}U or ^{232}Th should be used. As a consequence, the delayed neutron fraction is considerably reduced and the Doppler effect is small. The impact on safety parameters is strong and the requirement of a subcritical state, to assure inherent safety is desired. Maschek *et al.*, (2003) have calculated the kinetic parameters of a fuel mixture consisting of 25% Pu and 75% MA, for a 1200 MW_{th} core power at a subcritical level of $k_{eff} = 0.98$. Table 2.1 shows the deterioration of the kinetic parameters of the subcritical reactor in comparison with the Superphenix fast reactor.

Table 2.1 Comparison of fuel kinetics parameters between an ADSR and the Superphenix FR

Kinetic Parameters		ADSR	Superphenix
Doppler constant	[pcm/K][1]	-100	-860
β_{eff}	[pcm]	150	400
neutron generation time	[s]	$2x10^{-7}$	$4x10^{-7}$

The effect of these parameters on the reactor kinetics, can be studied through an analytical solution of the point kinetics equations for one delayed neutron group. In steady state conditions, the neutron density is proportional to the

[1] pcm = 0.00001 Δk/k

source intensity and the time constant is the prompt neutron generation time. The steady state neutron density is

$$n_0 = -\frac{q_0 l}{\rho_0} \tag{2.1}$$

where, n_0 is the neutron density, ρ_0 the initial reactivity, l the time constant of the prompt neutrons and q_0 the external neutron source per unit time. If one takes the ratio of two different steady states, the result would be proportional to the source change, and the initial reactivity as

$$\frac{n_1}{n_0} = \frac{q_1}{q_0}\left(\frac{\rho_0}{\rho_1}\right) \tag{2.2}$$

The new reactivity ρ_1 includes all possible reactivity feedbacks. This result indicates that any change (increase or decrease) in the source intensity, or a reactivity insertion leads to a power change. If for a moment, one neglects the possible reactivity feedbacks ($\rho_1 = \rho_0$), the result would indicate that the final power level during a transient is influenced only by the change of the source strength (Schikorr, 2001). On the other hand, if a positive or negative reactivity is inserted, the state of subcriticality will have an influence on the final power level.

In the subcritical reactor, the neutron flux has three contributors: the neutron source, the fissions and the delayed neutrons. The neutron source is independent of the fission process and is controlled by adjusting the proton beam current. The contribution of prompt and delayed neutrons to reactor kinetics is proportional to the initial neutron density, as well as to neutrons originating from the source. If there is a reactivity insertion, the neutron density initially increases by the prompt neutrons from new fissions (for $\rho_1<\beta$). Similarly, the delayed neutrons provide a fraction of this flux, sufficient to dominate the total flux increase. The reliability of the shutdown has to be redundant and independent, the system has to be capable of handling short beam trips without stopping the accelerator, but, in case of a large positive reactivity excursion, it has to stop the beam immediately. If the control system fails to stop the beam, the design must hold sufficient delay before unacceptable limits are reached, allowing one of the redundant systems to work. In practice, new measurement techniques are investigated to monitor the core reactivity and carry out control activities (Coddington *et al.*, 2004).

Reactivity Feedbacks

An ADSR does not respond to reactivity feedbacks as a critical reactor. The presence of the neutron source has an effect of reducing the sensitivity to reactivity changes, as it was explained through equation 2.2. On the other hand, the strength of the feedback effect depends on the specific fuel design and in particular, the choice of the subcriticality level. When the reactor is more subcritical, more importance is taken by the source and less effect is produced by the reactivity feedbacks. Therefore, to take advantage of the subcritical level, the subcriticality must be carefully balanced with the possible positive reactivity insertions and negative feedbacks. It has been argued that a rearrangement of fuel may lead to critical configurations of the core (Eriksson and Cahalan, 2002). It is therefore of primary concern to increase the Doppler effect in fertile-free fuel. For this purpose, Eriksson *et al.*, (2005) have studied inherent safety aspects of different fuels and concluded that, the higher melting point of Cermet fuel in combination with its larger critical mass hold the most favourable characteristic for the recriticality issue.

Another significant issue is the presence of high void worth of liquid metal cooled cores. This introduces a positive reactivity feedback that can exceed the Doppler effect. The positive void reactivity effect is a result of the fuel composition and the pitch to diameter ratio. The coolant-boiling problem could be relieved choosing a coolant with a high boiling point. Thus, voiding from coolant vapor formation is not likely to occur. Gas formation and entrainment in the coolant is an alternate path to voiding, it is also very unlikely since the sources of such gas entrainment are not sufficiently large and coherent to generate positive reactivity effects (Tang et al., 1978). Voiding as a consequence of fuel failure is a third mechanism of voiding. This can be the consequence of release of retained fission gas from the fuel pins or from local over heating from overheated fuel during postulated accidents. These sources are also very unlikely to produce voiding and to insert large amounts of positive reactivity. However, if a larger number of pins were to fail simultaneously then more serious consequences would occur. Then, core voiding could be presented during a large release of gas fission products accumulated in the fuel pin. Chen *et al.*, (2004) has computed different values of void coefficients for different type of fuels and it is seen that the Cermet fuel presents the lowest predicted value. This is caused by the smaller core size and the higher thermal conductivity that allows higher linear rating.

The pumping system has also to be studied against the problem of nuclear stability. If a large bubble passes through the core, a safety margin needs to be provided combining the coupled thermal-hydraulic and reactivity disturbance. The reactivity effects of the bubble will depend on the void worth throughout the core. If the coolant is lead alloy and for an equal gas mass, the bubble is going to be volumetrically smaller by as much as a factor of ten than in the case of sodium and therefore its reactivity effects are much smaller. For same volumetric bubble size, the consequences may be somewhat less because of the larger flow area used in naturally circulating lead and therefore less resistant to the bubble passage and a shorter transit time. In spite of these facts, to achieve a negative void worth in a liquid metal reactor requires the core design to either be very flat (core diameter much larger than the fuel height) or very tall (fuel height much larger than core diameter). Either configuration results in a high level of neutron leakage and a negative reactivity effect with the assumption of voiding. These approaches move the ADSR design away from the optimum economical configuration and accelerator power requirements. Detailed analysis of postulated events are necessary to support the risk assessment and similarly assess the probability of occurrence. It is desirable to avoid designs that would achieve prompt criticality from a postulated large bubble passing through the core.

Axial thermal expansion of fuel pins and radial thermal expansion of the core subassemblies lead to negative feedbacks. It is still a question whether the negative feedbacks are sufficient to control a reactor excursion, if the proton beam remains on. Other types of safety measures are necessary to assure the control of neutron source transients. The reactor safety strategy for the ADSR needs to be redefined with respect to the conventional fast and thermal reactors.

2.3 Thermalhydraulic Issues

Coolant Type

The coolant selection is a trade off between several aspects from chemical properties to economical and technical aspects. Table 2.2 gives an overview of some coolant characteristics and their effects on reactor safety. Now, if the timeframe for a selection is short, the choice is limited by proven technologies, such as sodium or lead/lead-bismuth (Spencer, 2002). In the case of long-term plans, the choice can be extended by the research and development tasks of various alternatives.

Table 2.2 Liquid metal coolant characteristics

Coolant	Advantages	Shortcomings
Sodium	Suitable neutronics	Activity with O_2 and H_2O
	Suitable thermalhydraulics	Gamma activity
	Not corrosive	Medium melting point
	Low radiotoxicity	
Lead-Bismuth	Suitable neutronics	Volatile alpha activity of Po
	Suitable thermalhydraulics	Highly corrosive
	No activity with O_2 and H_2O	Medium melting point
	No gamma activity	Long radiotoxicity
Lead	Suitable neutronics	Low alpha activity of Po
	Suitable thermalhydraulics	High corrosive
	No activity with O_2 and H_2O	High melting point
	No gamma activity	Long radiotoxicity

Lead-bismuth has been selected as the coolant for the ADSR primarily because of two criteria: performance and safety. LBE offers low moderation, low absorption cross-section, excellent heat transfer coefficients, high boiling point and low system pressure. The moderation and absorption cross section are important parameters determining the neutronic efficiency of the transmutation process. The low operating pressure reduces the structural requirements to the reactor vessel and the high boiling point minimizes the possibility of positive void reactivity insertions. The freezing point is an important issue for the coolant-structural interaction and potential failure by flow blockages. For this reason, lead-bismuth with a lower melting point (400 K) has been preferred as its counterpart "pure lead", which is cheaper but has higher melting point (600 K).

Other important issues regarding safety with LBE coolants are the activation and radiation damage. Lead-bismuth and also liquid metals do not dissociate or transform into different compounds, however, the main constraints are the long radiotoxicity of the ^{205}Pb and $^{210-208}Bi$ isotopes (over millions of years) and the presence of alpha activity originating from ^{210}Po, which is very volatile and increases the emissions at the free surface level of the reactor pool. Another disadvantage is found on the chemical compatibility with metals, which leads to corrosion and degradation of structures (IAEA-TECDOC-1289, 2002). Sodium, which is a strong candidate for liquid metal fast breeders, has been discarded for the ADSR because of its explosive reaction in contact with air or water. Regarding natural circulation capabilities, Ceballos *et al.*, (2004) have presented a study on liquid metal performance, demonstrating that the flow rate in natural convection systems is a function of the non-dimensional Stanton number and the Reynolds number. Some of these conclusions will be later exposed in chapter 4, when we analyze the steady-state reactor operation using dimensional analysis to clarify the nature of buoyancy flows.

Coolant Flow

To avoid loss of flow from a breakdown of the pumps, buoyancy driven cooling systems have been proposed (Rubbia *et al.*, 1995; Ansaldo, 2001). In these systems, the pumping power is determined by the pool configuration such as pool components and vessel height. From a safety perspective, the heat removal with natural circulation could be enhanced with gas lift pumps. This design has good safety characteristics and offers a bigger safety margin during transients (Cheng *et al.*, 2004).

The flow is subjected to a balancing system of buoyant forces and pressure drops. The pressure drops are proportional to the geometrical design of the flow channels and the flow velocity distribution. The core pressure drop is the largest portion of the total system pressure drop and therefore, the core height and diameter is always optimized for fuel economy and required pumping power. In the core, the flow has local characteristics since the power varies with respect to position. This leads to a local maximum fuel pin temperature profile that restrains the maximum core power. Another flow phenomenon is gas entrainment, since it can lead to reactivity insertions. The sources of entrainment could be fission gas release from a fuel pin rupture, a steam generator pipe break-up with carry-under or the gas entrainment from the gas covering the pool. The steam generator tube rupture event was identified as a potential source for extensive voiding and according to Eriksson *et al.*, (2005), the nitride and Cermet fuels hold a lower temperature peak during a positive void reactivity excursion. The gas cover entrainment has been studied and predicted for the liquid metal sodium reactor, surface waves and vortex formation by high velocities could result in encapsulation of bubbles carried deep below the coolant surface (Tang *et al.*, 1978).

The heat exchanger is a vital element for plant performance and safety. It links the radioactive and non-radioactive zones of the plant and also determines in great part the efficiency of the power cycle. The heat exchanger must permit a safe, stable and reliable operation under all conditions. For the case of a buoyancy cooled reactor, the design should minimize the pressure drop and provide good heat transfer performance. The decisions during design stage regarding secondary coolant type, flow configuration (in tube – in shell) and thermodynamic parameters (inlet - outlet temperature, pressure) are of relevance in the optimization of the natural circulation cooled reactor. The effect of partial flow blockages is important if the flow is obstructed and the cooling capability is lost resulting in a temperature rise and possible coolant boiling as well as cladding failure. Flow cavitation, which is caused by the reduced local pressure below the saturation vapor pressure, can induce vibration and erosion. Hydrodynamics studies of flow transitions have to demonstrate that this problem will not occur at any condition.

Core Power Limits

The maximum operating power level of the ADSR is limited by technical and economical requirements. The intensity of the neutron source necessary to drive the subcritical core depends on the spallation capacity to multiply the neutrons and the degree of subcriticality. The neutron multiplication increases with the increase of k_{eff} and also, with the increase of the proton beam energy, because the number of spallation neutrons is increased. Yamamoto and Shiroya (2003) have studied the performance of the neutron multiplication of 3 different target materials, depleted uranium (DU), lead (Pb) and Tungsten (W). The results have demonstrated a considerable impact on the neutron yield. DU provides the higher thermal flux, followed by an slightly higher thermal flux from Pb over W, but, Pb has a smaller absortion cross section than W, giving it a greater multiplication rate.

The primary cooling circuit can be either a forced or a natural circulation system. For this study, natural circulation is preferred because of the inherent passive safety that removes the possibility of loss of flow accidents. Davis *et al.*, (2002) have performed a thermal-hydraulic evaluation of an actinide burner cooled by natural circulation. The study has determined the maximum core power allowed optimizing various design parameters and minimizing the capital cost as well. Key technical issues in the design of a natural circulation cooled liquid metal reactor were identified as the heat exchanger type and tube lattice, the secondary cooling system pressure (for steam generators), the total system pressure drop and the vertical distance between the heat exchanger and core, which is limited by the vessel size. Besides, there are other types of technical limitations, which are intrinsic to all type of nuclear reactors such as the maximum fuel and cladding allowable temperatures to withstand thermal stress and strain. The removal of decay heat is another important factor to consider when limiting the maximum power for reactor operation. Decay heat should be transferred to an auxiliary cooling system when emergency conditions occur and the heat exchanger is not available. Some designs propose the decay heat removal by RVACS - Reactor Vessel Auxiliary Cooling System (Davis *et al*, 2002; Ansaldo, 2001). These operate transferring the heat from the fuel pins to the coolant and then, to the reactor vessel wall, which is cooled by natural convection of air. The radial power profile in the ADSR seems to have a large impact on the power restrictions, due to a higher power density observed around the central region. This effect is increased with the increased of subcriticality level (Rubbia *et al.*, 1995).

2.4 Mechanical Issues

The high neutron flux and the high temperature conditions place severe requirements on materials in the reactor core, particularly to the beam window and the fuel pin cladding. Both provide a containment barrier to radioactivity release. The fuel pins especially provide the basic structural integrity for the fuel elements. The probability of pin failure depends on the internal fuel pin gas pressure and the service conditions during lifetime. The cumulative damage evaluates the damage fraction during the steady and transient conditions for creep and fatigue.

The number of cycles to fatigue failure depends on the stress level and temperature history, the damage effect is usually measured as a reduction on the number of cycles. Fatigue considerations for other structural components exposed to the coolant flow are also important in the safety assessment. The flow distribution and mixing is important to reduce effects of thermal transients to structures. The material strength is dependent on the strain rate at high temperatures and if long time deformations persist, creep fluence occurs. Small strains with large hold times are more damaging than larger strain applied during short periods (Sauzay *et al.*, 2004).

The high coolant density of Pb-Bi introduces new issues in structural support and seismic response as well. The vessel height should be limited to reasonable margins in order to achieve a safe and cost effective design. MacDonald and

Buongiorno (2001) have limited the maximum vessel height to 19 meters because of transportability reasons, considering that it is an economical advantage if the reactor vessel can be built in a factory and being transported to the plant site. Davis *et al.*, (2002) consider that a higher vessel could reduce the cost adjustment over capital cost if there is an increase on electricity production. Metal degradation due to liquid metal exposure is another important factor in the reduction of mechanical properties of materials. Aiello *et al.*, (2004) have studied the effect of flowing Pb-Bi alloy on cladding steel T91. The tensile results shows a reduction of ductility and a fractured surface. Penetration of the liquid metal may be one of the reasons for the reduction of mechanical properties. In Russia, steel corrosion protection from Pb-Bi flows has been done mainly by maintaining a certain concentration of oxygen in the liquid metal, necessary to create a protective film of Fe_3O_4 on the steel surface (Ilincev, 2002). The irradiation effects are interpreted as irradiation hardening at lower temperatures and recovery at higher temperatures (Uwaba and Ukai, 2004). Irradiation hardening due to neutron displacement creates dislocations, which considerably affect the metal strength.

2.5 Design Preferences

Any reactor has to achieve the best thermal performance keeping some design parameters below certain limits. These limits have been gathered throughout fundamental analysis and experience. A list of design parameters and their limiting values for the ADSR have been introduced in table 2.3. The subdivision in groups relates the parameter to the physical system where it belongs. Due to our impossibility to perform experimental investigations and lack of design experience with liquid metal technology, we have appealed to the literature review, engineering sense and assumptions in order to draw some values.

First, the design has to overcome the limits relating the integrity loss by creep damage, which is consider as a percentage of strain during the service time, the values are equal to 1% limit at individual points, 2% at the central line of a structural cross section and 5% when averaging the whole cross section. In the same way, the design criteria states that the maximum allowable stress at any point should not exceed the yield shear stress of the uniaxial strength test. Additionally, to limit corrosion, erosion, cavitation and vibration effects, the velocity has been limited to 1 - 1.5 m/s. The choices with regards to thermalhydraulic effects are related to temperature limits to avoid creep damage of structures as well as fuel and cladding melting. Hence, values for cladding, structures and fuel are 650 C, 450 C and the fuel melting temperature respectively. Limiting temperatures based on accelerated corrosion effects have been excluded since there was not literature available for this case. Finally, the choice of a suitable subcriticality level is limited to $0.95 < k_{eff} < 0.98$. This range considers sufficient safety margins to prompt critical excursions as well as accelerator power availability.

Table 2.3 Design limits for the ADSR

Constraints		Design limits	Reference
Structural	Creep strain	1% local	ASME
		2% in the central line of the cross section	ASME
		5% in the average cross section	ASME
	Fatigue	$\sigma < \tau_{yield}$	ASME
	Coolant velocity	< 1 - 1.5 m/s corrosion - erosion	Fomichenko
		< 9 m/s cavitation and vibration	Tang
Thermal-hydraulic	Cladding	Temperature limit normal operation < 650 C	Davis
	Fuel	Fuel temperature < melting temp.	Davis
		overpower conditions 115%	Tang
	Reactor	Vessel temp. steady < 450 C	MacDonald
Neutronics	Kinetics	subcriticality level - 0.95 < keff < 0.98	Rubbia
	Reactivity coef.	Doppler coefficient < 0	

2.6 Safety Constraints

Having identified the safety concerns and design conditions of the ADSR, one has to establish limits for the safety related parameters for transient conditions. These limits are summarized in table 2.4 and together with table 2.3 form the safety envelope introduced in figure 2.1, at the introductory section of this chapter. As it was outlined, if one or more of these limits are exceeded during steady or transient conditions, the controllability and/or integrity are exposed. Safety analysis should ensure that at any scenario, the reactor parameters will remain within the safety envelope. The new parameter integrating the safety constraints for the structural systems is the cumulative damage, which measures the effect of cycles on the lifetime of components. The damage is a fraction of the total life predicted with experimental investigations for similar conditions. This concept will be later explained in chapter 6 when we deal with the structural integrity assessment of the cladding. The thermalhydraulic system considers cladding temperature limits to avoid creep. In the same way, the fuel temperature is limited to avoid fuel melting and the coolant is limited to its boiling temperature; the last in order to avoid positive reactivity insertions. Finally, the reactivity is required to be always negative in order to remain subcritical, any positive excursion that brings the reactor critical or prompt critical is considered as a major loss of control and delivers high risk of accidents.

Table 2.4 Safety limits for the ADSR

Constraints		Safety limits	Reference
Structural	Creep strain	1% local	ASME
		2% in the central line of the cross section	ASME
		5% in the average cross section	ASME
		Cumulative Damage < 1.0	ASME
	Fatigue	$\sigma < \tau_{yield}$	ASME
		Cumulative Damage < 1.0	ASME
	Creep-Fatigue	Cumulative Damage < D	ASME
	coolant velocity	< 1 - 1.5 m/s corrosion - erosion	Fomichenko
		< 9 m/s cavitation and vibration	Tang
Thermal-Hydraulics	Cladding	Temperature limit transients < 750 C	Davis
		Temp. limit anticipated transient <788 C	Tang
		Temp. limit unlikely events transient < 871 C	Tang
	Fuel	Fuel temperature < melting temp.	Davis
		overpower conditions 110%	Tang
	Reactor	Vessel temp. transients < 750 C	MacDonald
	Coolant	Coolant Temp. < Boiling temperature	MacDonald
Neutronics	Kinetics	Reactivity < 0	

2.7 Safety Analysis

Following the conventional approach for nuclear plant safety assessment, hypothetical scenarios risking the reactor integrity and controllability are studied to ensure that the safety limits are not exceeded. Different design codes categorize these events in similar way, therefore, we have decided to follow the ASME code classification, which advises to divide the possible scenarios in normal, upset, emergency and faulted events. Table 2.5 presents a list of events for an ADSR safety assessment. Normal events are the operating conditions like steady state and routinely transients, normal start-up and shutdown. Upset events are anticipated faults like incidents which does not cause any repair consequence or damage, they are expected to occur once or more during the lifetime of the plant. For an ADSR, the beam trips and the loss of heat sink consequence of a failure in the secondary cooling circuit are expected events. Emergency events are unlikely faults requiring the shutdown and repair activities, it may involved the reduction of lifetime of components, but, not the break down of structures. In this class, the flow blockages or positive reactivity insertions could be fitted. Faulted events are major incidents of very low probability but involving the risk of complete structural integrity loss, requiring repair activities or extensive inspection. They are large reactivity insertions or the unprotected transients where the control fails in shutting down the proton beam. Analysis of these transients will be later conducted along chapters 4 and 5.

Table 2.5 Classification of scenarios for the safety analysis of the ADSR

TYPE	Initiating event	Definition	Probability (event/year)
Normal	Full power	Normal operation	1
	Startup	Startup	
	Shutdown	Shutdown	
	Proton beam fluctuations	Beam interruptions	Several
Upset	Decrease of heat removal from heat exchanger	partial LOHS	10^{-2}
	Secondary cooling system pump failure	LOHS	
	Loss of electrical power in secondary cooling	LOHS	
	Proton beam fluctuations	Beam interruptions	
	Loss of electrical power in accelerator building	Beam trip	
	Loss of electrical power in reactor building	LOHS	
	Partial loss of enhanced gas lift with beam shutdown	partial LOG	
	Total loss of enhanced gas lift with beam shutdown	LOG	
Emergency	Fuel assembly partial blockage	Partial Blockage	10^{-2} - 10^{-4}
	small pipe break in heat exchanger and steam carry under	Positive reactivity	
	Small pin failure with gas release	Positive reactivity	
	Proton beam fluctuations	Beam interruptions	
Faulted	Large pipe break in heat exchanger	Positive reactivity	10^{-4} x 10^{-7}
	Large pin failure with gas release	Positive reactivity	
	Loss of enhanced gas flow without beam shutdown	ULOG	
	Loss of electrical power in secondary cooling system without beam shutdown	ULOHS	

Maschek *et al.*, (2003) have studied ADSR accident scenarios and proposed a system of 3 shutdown levels, which follows the lines of defense concept. The first shutdown level corresponds to the neutron source shutdown, which it will rely only on monitoring and detection systems. A second shutdown level requires the insertion of additional absorber rods. This shutdown level resembles a redundant shutdown system. A third shutdown level relies on inherent and passive measures from the design like selecting the core geometry with favorable reactivity coefficients and assuring reasonable kinetic parameters. The kinetics characteristics of ADSR presented in section 2.2 have demonstrated that it is necessary to manage the neutron source in order to achieve inherent shutdown. Ericksson and Cahalan (2002) have studied the inherent shutdown based only on reactivity feedbacks and proved that it is unfruitful. They have determined that the fastest transients are the overpower transients caused by a maximum insertion of the beam power. Transients without beam shutdown have a severe impact on integrity and some shutdown device is required, therefore, they have proposed some concepts to achieve the passive ADSR shutdown.

A safety strategy that relies on redundant shutdown systems is not sufficient to assure that they will handle all possible transient scenarios. Thus, the implementation of passive safety mechanisms should be at the heart of an integral safety strategy for the ADSR. The development of a reactor safety strategy for the ADSR is especially needed since it is a new innovative and not well-known combination of different technologies.

2.8 Concluding Remarks

The present chapter has introduced some of the more important safety related issues for the ADSR. The identification of relevant physical parameters is important to establish the safety envelope for the reactor operation. For design and safety analysis, one has to constraint these parameters. This is supported by literature review, engineering sense and some assumptions. In addition, the categorization of scenarios gives the structure to the safety analysis work. The performance required will be measured according to the safety limits and the results will contribute to specify the requirements for a reactor safety strategy. The implementation of passive safety principles is at the heart of an integral safety strategy and that is the aim of the present work.

"If you prick us, do we not bleed? If you tickle us, do we not laugh? If you poison us, do we not die? and if you wrong us, shall we not revenge?"
The Merchant of Venice, W. Shakespeare, 1564-1616

3 Physical Modeling of the ADSR

This chapter presents a description of the physical models used for the simulation of the ADSR under steady-state and transient conditions. The model consists of three different sub-models: a reactor kinetics model, based on simple point kinetics theory including an external source, a fuel pin model and a thermalhydraulics model of the reactor pool. The latter model evaluates the natural convection capabilities of the coolant with single-phase and two-phase flow for 1D and 2D axisymmetrical geometries.

3.1 Introduction

During the fission process, energy is deposited in the fuel as heat. The heat conducts through the fuel pin and is transferred to the coolant flowing outside the fuel pin via convection. Figure 3.1 shows the fuel pin temperature gradient as function of the pin radius. As the thermal conductivity is different for the different materials and heat is only generated in the fuel, the temperature profile changes from the fuel zone to the gap and cladding.

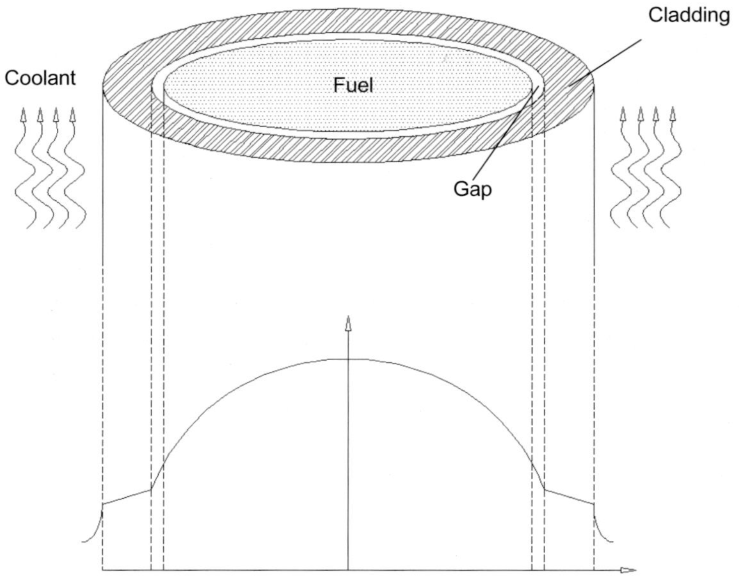

Figure 3.1 Temperature profile as a function of the radius in a fuel pin with an internal heat generation given from the nuclear fission process

The reactor core is formed by more than one hundred subassemblies each containing dozens of fuel pins. Figure 3.2 shows the fuel pins arrangement in a single fuel assembly. The subassembly concept was proposed by MacDonald and Buongiorno, (2001) and used by Davis *et al.*, (2002) in the design of a natural circulation lead-bismuth cooled actinide burner. The fuel assembly design is very particular because it holds a small pitch/diameter ratio to reduce void worth and uses gas-filled tubes at the centre and the periphery of each fuel assembly to promote neutron leakage exhibiting a favorable coolant void worth.

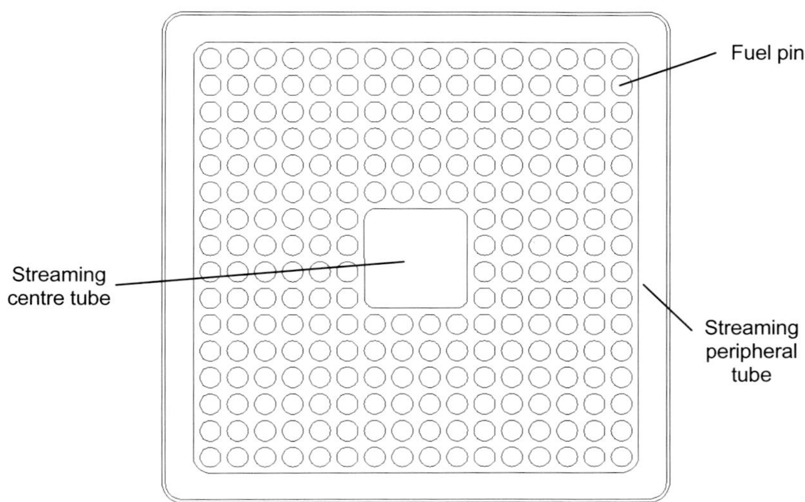

Figure 3.2 Core top view with fuel assemblies distribution (at the left) and fuel assembly design (at the right)

The fuel pins are surrounded by the coolant, which flows in a closed circuit inside the reactor pool. The heat released from the fuel pins in the core heats up the liquid metal, the fluid density decreases and the fluid moves upward. The total buoyancy force is given by density differences between the riser and the downcomer. Figure 3.3 shows a scheme of the reactor pool. It displays the different regions like core, heat exchanger, riser and downcomer. It also includes the power cycle and the gas lift pump system. The gas injected inside the riser, is meant to decrease the fluid density and produce additional buoyancy. The heat exchanger removes the heat from the reactor pool and releases its energy into a conventional power cycle.

The reactor model consists of three different sub models: the nuclear kinetics model, the fuel pin heat transfer model and the thermalhydraulics model of the reactor pool. The nuclear kinetics model applies point kinetics theory with 6 delayed neutron groups and a decay heat model with 23 decay time groups. The model predicts the power evolution and defines the heat source in the fuel. The fuel pin heat transfer model calculates the temperature profile inside the pins considering the internal heat generation from fission and decay. The resulting temperature profile determines the averaged fuel temperature for the neutron kinetics model to estimate reactivity changes.

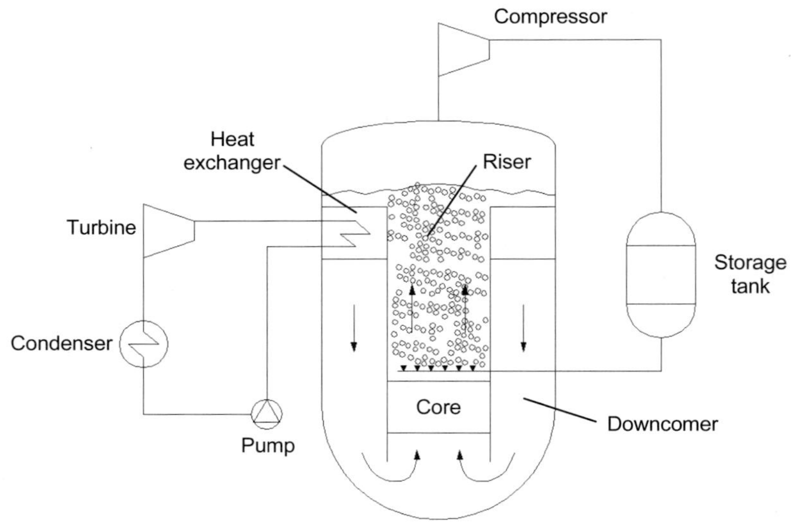

Figure 3.3 Reactor pool with power cycle and coolant flow scheme including the gas lift pump system

The thermalhydraulics model describes the coolant flow in different regions like the core, riser, heat exchangers and downcomer. Its mathematical formulation is based on the momentum and energy conservation laws for a closed loop. The model considers the momentum source provided by buoyancy forces from thermal gradients and two-phase flow when gas is injected in the riser. The thermalhydraulics model contains three submodels: the heat exchanger, the core and the riser model. It solves the coolant temperature and mass flow rate in the reactor pool given the heat transferred from the fuel pins to the coolant and heat removed by the heat exchanger. The set up of these models is explained graphically in figure 3.4.

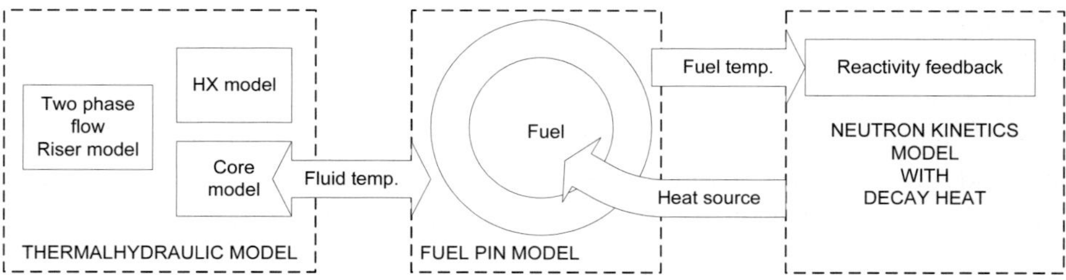

Figure 3.4 The reactor dynamic model and its 3 physical sub-models: the thermal-hydraulic, the fuel pin and the neutron kinetics

3.2 Neutronic Model

Introduction

The computation of neutron kinetics is usually obtained by solving the neutron transport equation. The direct solution of the spatial, energy, time dependent equation can be achieved with advanced numerical schemes that provide sufficient accuracy although the computational work is "expensive". A simpler solution is usually obtained with the exact point kinetics approximation which writes the solution as a product of a spatial function and a time dependent function. The solution method calculates the spatial shape less frequently than the time dynamic response (Ott and Neuhold, 1985).

An alternative solution proposed by Rineiski and Maschek (2005) for the subcritical reactor keeps a single time-independent weighting function, but employs a variable flux shape. The model offers a table of power shapes computed in the k_{eff} mode for the initial conditions and several shapes computed at different reactivity levels. This approach, can improve significantly the speed of the calculations, although perfect agreement with exact point kinetics is never achieved.

The derivation of a simpler point kinetics model assumes that the flux shape never changes in time. Many safety codes apply this simpler and faster point kinetics model for studying transients of critical reactors. The computation of parameters such as reactivity and reactivity coefficients are integrals over the space and can be determined a priori. Eriksson *et al.*, (2005) have investigated the ability of simple point kinetics to predict the transient behavior of an ADSR. Numerical experiments were carried out to compare the precision of point kinetics against the full solution of space-time dependent kinetics. The results suggested that point kinetics is capable of accurate solutions for transients involving external source perturbations with small flux deformations, because of the uniformity of the reactivity feedbacks. When the transient involves localized perturbations, the results indicates better precision if k_{eff} is lower (more subcritical). This behavior is due to the lower sensitivity to the feedback effects and the major role played by the source on the power output.

If the system is near critical, the neutron distribution is close to that of the fundamental eigenmode. But, when the reactor is far subcritical, the deviation can be significant and the distribution will depend not only of the fundamental mode, but also on other excited modes (Rydin and Woosley, 1997). The principal condition for accurate point kinetics modeling in subcritical systems is therefore, the use of an optimal weighting method to calculate the kinetic parameters.

For our safety analysis, it is an advantage to make use of simple point kinetics considering the lower computational cost and the relatively good accuracy. Nevertheless, one must be aware that the conditions to apply point kinetics in subcritical systems are the same of critical reactors: uniform reactivity insertions and a highly coupled core. Transient scenarios that involve local effects should apply the spatial-time kinetics method.

Model Description

The point kinetics model calculates the dynamic response of the reactor to reactivity changes and external source perturbations. The model computes both the fission power and decay power. Six groups of neutron precursors and a decay heat model with 23 time groups according to the ANS standard (1979) are part of the system of equations. The mathematical formulation of point kinetics including decay heat is

$$\frac{dP(t)}{dt} = \frac{\rho(t)-\beta}{\Lambda}P(t) + \sum_{i=1}^{6} \lambda_i C_i(t) + \frac{S(t)}{\Lambda} \tag{3.1}$$

$$\frac{dC_i(t)}{dt} = \frac{\beta_i}{\Lambda}P(t) - \lambda_i C_i(t) \tag{3.2}$$

$$\frac{dP_i(t)}{dt} = \frac{\gamma_i}{\lambda_{i\gamma}}P(t) - \lambda_{i\gamma} P_i(t) \tag{3.3}$$

$$P_t(t) = P(t) + G(t)\sum_{i=1}^{23} P_i(t) \tag{3.4}$$

where *P(t)* is the prompt power, *C(t)* is the precursor concentration expressed in power units, $\rho(t)$ is the total reactivity, β the effective delayed neutron fraction, β_i the delayed neutron fraction from delayed group *i*, Λ the neutron generation time, λ_i the time constant of delayed group *i*, *S(t)* the external source, $P_i(t)$ the decay power from fission products, γ_i and $\lambda_{i\gamma}$ fitted parameters from data of fission decay power and $G(t)$ a correction factor that accounts for effects from neutron absorption, not considered here assuming that the reactor has been loaded with fresh fuel.

Appendix 1 lists the values used for the solution of the point kinetics model. The delayed neutron precursor fractions and time constants were taken from the benchmarking exercise of beam interruptions in ADSR (D'Angelo and Gabrielli, 2003). The decay heat quantities are assumed to be the same as those of ^{235}U. The Doppler reactivity feedback is obtained according to

$$\rho(t) = \rho_o + \alpha_D\left(\left\langle T_f \right\rangle - T_o\right) \tag{3.5}$$

where α_D is the Doppler reactivity feedback coefficient, $\langle T_f \rangle$ is the average fuel temperature and ρ_o, T_o are the reactivity and fuel temperature reference values.

3.3 Fuel Pin Heat Transfer Model

The following assumptions were used to develop a heat transfer model for the fuel pins:

- The heat transfer in the fuel pin is given only by radial heat conduction through the fuel, gap and cladding
- The thermal conductivity of the fuel, gap and cladding is assumed to be independent of temperature
- Transient heat transfer from fuel pins to liquid is assumed to follow the correlations for steady state heat transfer
- The volumetric heat source in the fuel pin is constant in radial direction
- Fuel pin spacers are small and can be disregarded in the determination of the flow cross sectional area

The fuel pin heat transfer is governed by the heat conduction equation in cylindrical coordinates including an inner heat source and neglecting axial heat conduction as

$$\rho C_p \frac{\partial T}{\partial t} + \frac{1}{r}\frac{\partial}{\partial r} rk \frac{\partial T}{\partial r} = Q \tag{3.6}$$

with boundary conditions

$$\left.\frac{\partial T}{\partial r}\right|_{r=0} = 0 \qquad \left. k\frac{\partial T}{\partial r}\right|_{r=R} = q'' \tag{3.7}$$

where ρ is the density of the material, C_p the heat capacity, r the radial coordinate, k the thermal conductivity, Q the heat source term resulting from nuclear fission or decay, R the external radius of the cladding and q'' the heat flux per unit area.

The convection heat transfer coefficient used to calculate the heat flux q'' is obtained with Nusselt number correlations for liquid metal heat transfer. The equations are fitted to fully developed steady-state heat transfer experiments for uniformly heated fuel bundles. Experimental and analytical studies of heat transfer in Hg, PbBi, NaK and Na coolants from several authors are summarized in a technical document from IAEA-TECDOC-1289 (2002). According to this comparison, the best fitting results for square array bundles with spacers are obtained using the Nusselt number proposed by Zhukov *et al.*, (2000)

$$Nu = 7.55\frac{p}{d} - 14\left(\frac{p}{d}\right)^{-5} - 0.09Pe^{\left(0.64 + 0.264\frac{p}{d}\right)} \tag{3.8}$$

where p is the pitch between pins, d is the diameter of the fuel pins and Pe is the Peclet number.

3.4 Thermalhydraulics Models

Introduction

The solution of the velocity and temperature distribution of the liquid metal flow is obtained by solving the mass, momentum and energy balances of the reactor pool. The flow analysis is performed with 1D and 2D axisymmetrical modeling. Both approaches solve flow parameters, which are influenced by characteristic pressure drops and heat sources in every region. The pressure drops and heat transfer coefficients are empirically defined relations in macroscopic scale of single-phase and two-phase flow. An important remark is that the physical properties of the coolant such as viscosity, thermal conductivity, heat capacity, thermal expansion and surface tension are all assumed to be temperature independent. It is a reasonable assumption for many practical problems of liquid flow and simplifies the mathematical treatment of the equations. It is also supported by the fact that there is only a small change of values for the operational range of the problem (600-900 K)

(Morita *et al.*, 2004; Alchagirov *et al.*, 2000). The coolant density is the only property made temperature dependent using the Boussinesq approximation.

1D Thermalhydraulics Model

Figure 3.5 shows a schematic representation of the natural circulation cooled reactor as a closed loop with different regions. The formulation of mass, momentum and energy balances describe the fluid flow in this loop. The following assumptions were used to simplify the model:

- The Boussinesq approximation for natural convection is valid
- Conduction of energy along the loop is neglected
- The flow regime is turbulent
- The heat supplied in the core is constant along the core height
- The riser and downcomer are considered adiabatic, unless a reactor vessel auxiliary cooling systems (RVACS) is specified, which means that heat transferred is proportional to the difference between the fluid temperature and the ambient temperature outside the vessel

If gas is injected in the riser, the following assumptions are used as well:

- There is thermal equilibrium between the two phases in the riser
- All gas injected at the bottom of the riser is removed at the top and there is no carry-under in the heat exchangers, downcomer and core

The momentum equation solved for the mass flow rate w, has followed similar steps as those presented by Todreas and Kazimi (1990) for a forced circulation reactor. The flow rate is given by

$$\begin{aligned}&\frac{dw}{dt}\sum_{n}\frac{l_n}{A_n}+\int_{riser}\left(\rho_{mix}(z)-\rho_{ref}\right)g\,dz+\oint\rho_{ref}\,g\,\beta_T\left(T(z)-T_{ref}\right)dz+\\&\frac{w^2}{2\rho_{ref}}\left(\sum_{n}\frac{K_n}{A_n^2}\right)+\frac{w^2}{2\rho_{ref}}\left(\sum_{n=1}^{N-1}f_n\frac{l_n}{D_n}\frac{1}{A_n^2}\right)+\frac{w^2}{2\rho_{ref}}\left(\phi_{fo}^2\cdot f_{riser}\frac{l_{riser}}{D_{riser}}\frac{1}{A_{riser}^2}\right)=0\end{aligned} \tag{3.9}$$

where n is the subscript that indicates the different regions of the loop like riser, heat exchanger, downcomer and core, t is the time, l_n the length of different regions, A_n the cross sectional area, g the gravitational constant, $\rho_{mix}(z)$ is the density of the gas-liquid mixture in the riser, β_T the thermal expansion coefficient of the fluid, $T(z)$ the temperature of the fluid along the loop, T_{ref} is the reference temperature of the fluid, ρ_{ref} the fluid density referenced at the reference temperature, z is the axial direction, K_n is the pressure drop coefficients for sudden contractions and expansions of pipes, f_n the friction factor, D_n the hydraulic diameter of the region, Φ_{lo} is the two phase friction multiplier, determined according to the Friedel model (Collier and Thome, 1998).

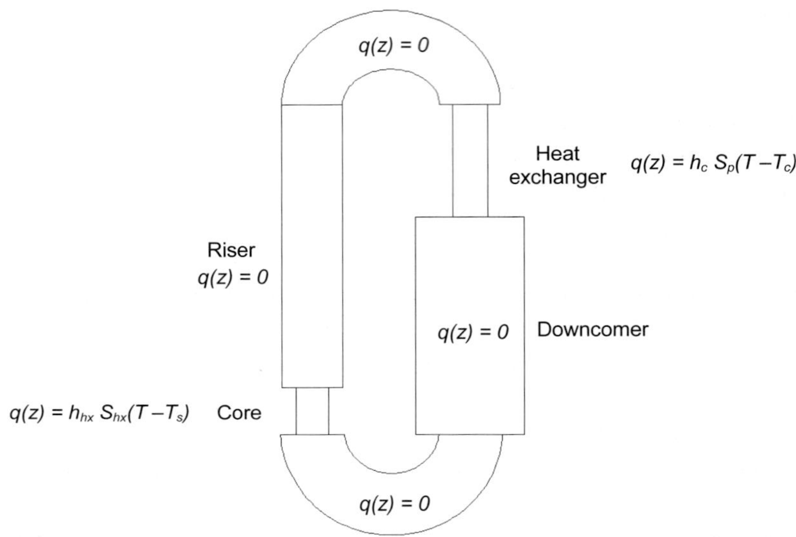

Figure 3.5 Schematic representation of the reactor pool as a closed loop

The energy equation is defined for every region as

$$C_p\left(\rho_{ref} A\frac{\partial T}{\partial t}+w\frac{\partial T}{\partial z}\right)=q(z) \tag{3.10}$$

where $q(z)$ the linear heat flux defined for every region as:

$q(z) = 0$	for the riser
$q(z) = h_c\, S_p\, (T - T_c)$	for the core
$q(z) = h_{hx}\, S_{hx}\, (T - T_s)$	for the heat exchanger
$q(z) = 0$	for an adiabatic downcomer
$q(z) = h_v\, S_v\, (T - T_a)$	for the downcomer with RVACS

h_{hx} is the heat transfer coefficient of the heat exchanger, h_c is the heat transfer coefficient of the fuel pins in the core, S_p is the perimeter of all pins outer diameter, S_{hx} is the perimeter of the heat exchanger pipes outer diameter, T_c is the cladding surface temperature, T_s is the heat exchanger pipe surface temperature, h_v is the heat transfer coefficient of the vessel, T_a is the ambient temperature of the vessel surroundings and S_v is the perimeter of the vessel.

2D Axisymmetric Thermalhydraulics Model

The model is built into the commercial CFD software Fluent. The advantage of using CFD is foreseen in the multidimensionality and robustness of the solver compared to the 1D model. Two-dimensional axisymmetric simulations are able to reproduce spatial details and consider the effect of turbulence. All assumptions from the 1D model are still valid. The CFD model solves the flow using the numerical grid representing the reactor geometry, which is shown in figure 3.6.

The model consists of a numerical domain divided in tetrahedral cells that represents the 2D axisymmetric geometry. The boundary conditions used for the solution of the flow problem consider all the walls as no-slip walls and the free surface of the pool as a free-slip wall. The heat transferred from the vessel to the surroundings is roughly estimated with a total heat transfer coefficient and the outside ambient temperature.

The core and heat exchanger regions are considered as zones with heat and momentum sources. The heat source from the core and heat exchanger are proportional to the temperature difference between fluid and wall. The proportionality constant is the convection heat transfer coefficient and the contact surface area. Their sources are implemented in the source term of the energy equation from the numerical cells of the CFD model. The core and heat exchanger pressure drops are modeled as porous regions with frictional losses.

The core heat source model contains subroutines to calculate the neutron kinetics model and the fuel pin model described in section 3.2 and 3.3. To capture the effect of the spatial power distribution, the core has been divided in different radial and axial zones. A matrix of weighting factors determines the power delivered in every specific zone of the core. The power distribution shape is taken from Rubbia *et al.*, (1995). There, the weighting factors are distributed in an arrangement of 20 radial zones by 10 axial zones. This arrangement was considered computationally too expensive and during the benchmarking stage of the model, 6 radial by 4 axial zones were found to give a good description within a reasonable computational time. The heat source model and heat sink model are implemented by means of user-defined functions (UDF) in the C++ language. The UDFs are compiled and linked to the CFD solver. Additional subroutines are added for the initialization of variables, adjustment of values at the end of the time step and initialization of transients. More details are presented below when the coupling strategy between submodel is explained.

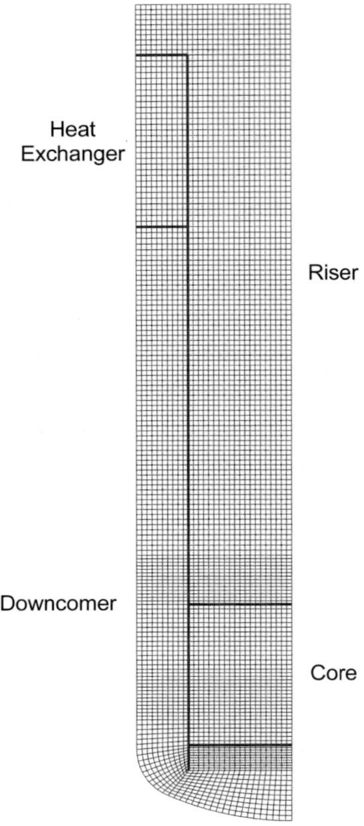

Figure 3.6 Grid used by the 2D-axisymmetric CFD model

Equation 3.11 describes the way these weighting factors are related to the power of each zone. Figure 3.7 emphasizes the difference between cells and zones: cells constitute the geometry subdivision for the flow computation whereas, zones are collection of cells with an equal weighting power factor and the same volumetric heat source given by the fuel pin model.

$$P_{zone} = PD_{avg} \cdot m_{fuel\_rod} \cdot N_{rod\_zone} \cdot WF \tag{3.11}$$

where P_{zone} is the power in a specified zone, PD_{avg} is the average power density of the core, $m_{fuel\text{-}rod}$ is the mass of the fuel in one rod, $N_{rod\text{-}zone}$ is the number of fuel rods in one zone and WF is the weighting factor of the power shape distribution matrix.

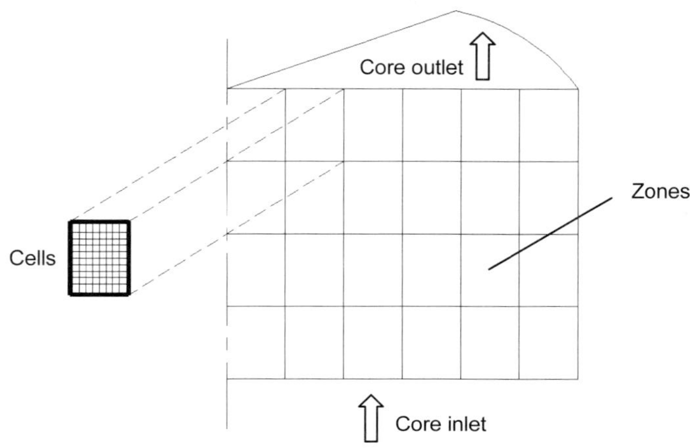

Figure 3.7 Description of the core zoning method to represent power shapes in the 2D axisymmetrical model

Fluent uses a finite volume method to transform the non-linear governing equations in a linear system of equations for the dependent variables at every computational cell. By default, Fluent stores discrete values of the scalar at the cell centres. Fluent uses a Gauss-Seidel linear equation solver together with an algebraic multigrid (AMG) method to solve the resultant scalar system of equations for the dependent variable in each cell.

The steps followed during every iteration are:

1. Fluid properties are updated based on the current solution or initialization
2. Momentum equations are solved using current values
3. Continuity equation is solved and if mass is not conserved, a pressure correction method from the continuity and momentum equations is used
4. Energy equations are solved
5. Turbulence scalars are solved
6. The source terms are updated for the next iteration
7. Check for convergence

The CFD solver facilitates the implementation of user-defined functions at different stages of the calculating procedure. Figure 3.8 shows the structure of the program that couples the neutronics model with the CFD thermalhydraulics model built in Fluent. This program is implemented in C++ language and linked to the solver with a shared library. The structure is divided in three main functions, which can be called from the fluent platform by means of predefined macros. The first function follows the 5th step, it reads data from CFD, adjusts values in the subroutine and solves temperatures of the fuel pins. Then, a second function defines heat sources for the core and momentum sources. Finally, the third function stores and writes the data needed for further processes.

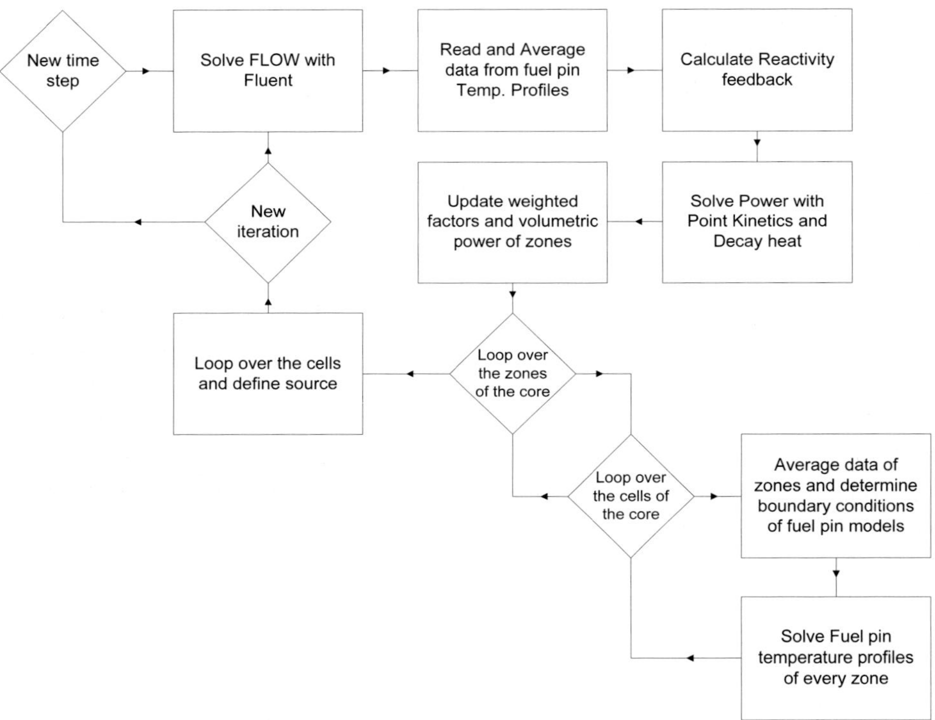

Figure 3.8 Structure of the algorithm for the interface of neutronics and thermalhydraulics

Figure 3.8 explains schematically what happens during a time step when the UDFs are called. First the mass flow, velocity and temperature distribution in the reactor pool are solved. Subsequently, the Doppler reactivity feedback is evaluated and the point kinetics equations with decay heat are solved. The weighting factors defining the power distribution shape are used to calculate the power density in the fuel pins. The program loops over the zones to determine fluid parameters and solve the temperature profile of the fuel pins. The approach used for averaging cell values in different zones converts the small scale from CFD cells into a bigger scale defined by the zoning. Then, the fuel pin temperature profile is obtained and the new values for the cladding surface temperature are used to assign a new heat source of the fluid cells. Finally, another function swaps data from new values to old values in order to be used during the next iteration and some other data are written for external recording.

The algorithm used for the heat exchanger model loops over the cells in the heat exchanger zone and adds a source term to the cells. The source term is defined as a temperature difference between the fluid (dependent variable from Fluent) and a given temperature, which corresponds to the heat exchanger pipe surface temperature, which is assumed constant.

Pressure Drop Model

Two main contributions determine the total pressure losses in the reactor pool. They are the friction losses of the fluid in contact with the walls and form losses at locations of changing cross sectional area, i.e. entering and leaving the core and heat exchangers. The friction coefficients are based on empirical correlations obtained through experiments at steady state conditions. These expressions are assumed to be valid for the transient conditions as well. Pressure drops are defined as the product of a dynamic pressure head and a loss factor given as

$$F = K_{form} \frac{\rho v^2}{2} + K_{\mu} \frac{\rho v^2}{2} \tag{3.12}$$

the form factor K_{form} is obtained from the contraction and expansion relations of smooth transition in pipes (Todreas and Kazimi, 1990). The friction factor K_{μ} is proportional to the friction coefficient f_i, the channel length and inversely proportional to the hydraulic diameter D_H. The friction coefficient depends on the channel geometry and the average velocity in the channel. Choi *et al.*, (2003) have measured the pressure drop of fuel pin assemblies in liquid metal flow. The result has shown that the proposed correlation by Cheng and Todreas (1986) fits best the experimental data and should be used for development in liquid metal flow analysis. To estimate the pressure drops in the heat exchanger, riser and downcomer, the friction coefficient expression is taken from Rust (1979). This expression was experimentally validated and has proved to provide good accuracy for a large range of Reynolds numbers flowing in circular pipes without restrictions (Garland *et al.*, 1998).

Heat Exchanger Model

It is not our purpose to provide detailed information about the power cycle behavior during transients. Hence, a simplified model of the heat exchanger is sufficient to account for the effect of this component on the thermalhydraulics system performance. The model considers that the total heat transferred across the heat exchanger pipes, should be proportional to the overall heat transfer coefficient of the heat exchanger U, the heat exchanger pipes surface area A, and the temperature difference between fluid and wall

$$Q_{hx} = UA(\Delta T) \tag{3.13}$$

Before the model is implemented, the heat transfer coefficient of the heat exchanger was calculated iteratively. The steady state heat transfer was matched to the heat exchanger energy balance. The calculations of the overall heat transfer coefficient considered the convective heat transfer outside and inside the pipe, the heat conduction in the oxide layer over the pipe and inside and the heat conduction across the pipe wall.

Two-Phase Flow Model

The gas is injected at the bottom of the riser and could come either as a bubble or as a turbulent jet depending upon the gas flow rate. When the flow rate is low, the bubbles form a vertical column that ascends with swirling motion. They could disintegrate by the shear forces into smaller ones of random size distribution (Satyamurthy *et al.*, 1998). The purpose of gas injection is to establish a homogenous gas-liquid mixture that enhances the buoyancy of the liquid metal flow. Therefore, the two-phase flow pattern in the riser should be kept in the dispersed bubbly flow regime thereby ensuring the homogeneity and avoiding possible flow instabilities and vibrations. In this way, a model has been defined exclusively for bubbly flow regimes and thus the void fraction limited to the maximum amount of void to keep the bubbly flow regime along the riser, this value equals 0.3.

Modeling of the two-phase gas-liquid metal flow has been performed with a multi-fluid model by Satyamurthy *et al.*, (1998) and Suzuki *et al.*, (2003). The approach used in both studies is the coupling of the momentum conservation equations of the two phases, considering the drag force between them. For our purpose, a simplified model that evaluates the flow enhancement by gas injection is sufficient to predict its effect on the reactor mass flow rate. The two-component systems are represented as a single homogeneous mixture. Drift flux correlations are used to estimate local void fractions for calculating the mixture density at different positions of the riser. This, because the bubble size changes as it moves upwards and expands due to hydrostatic pressure changes.

Model Description

When gas is injected, a gas-liquid mixture is produced inside the riser. The mass, momentum and energy balances for the riser use a mixture density, which is a function of the vertical position. The mixture density is proportional to the gas and liquid volume fractions and fluid densities as

$$\rho_{mix}(z)=\alpha(z)\rho_g(z)+(1-\alpha(z))\rho_{ref} \tag{3.14}$$

where $\rho_g(z)$ is the gas density function of the gas temperature and hydrostatic pressure and $\alpha(z)$ the void fraction at different positions of the riser.

The main assumptions for this mixture model are:

- The two phases are considered to be in thermodynamic equilibrium
- The mass transfer between phases is neglected
- The gas is assumed to be in quasi-equilibrium with the hydrostatic pressure of the liquid and its density is calculated from the ideal gas law

The drift flux model calculates the void fraction as function of the superficial velocity j_g^+, and the drift flux V_{gj}^+. The cross sectional averaged void fraction is obtained from

$$\alpha(z)=\frac{j_g^+}{C_o j_g^+ + V_{gj}^+} \tag{3.15}$$

where C_o is the distribution parameter, given by the expression for round tubes

$$C_o = 1.2 - 0.2\sqrt{\frac{\rho_g}{\rho_f}} \tag{3.16}$$

The terms with the superscript $^+$ are normalized by the quantity $(\sigma g \Delta\rho/\rho_{ref})^{0.25}$, where σ is the surface tension of the fluid and $\Delta\rho$ is the difference of density between liquid and gas.

The flow pattern selection is taken from a dispersed bubble flow map given by Clift *et al.*, (1978). According to this map, the flow regimes for liquid metal two-phase flows are limited to the ellipsoidal and cap bubble type, because of the low Morton numbers caused by the large surface tension. The limiting value for the change from ellipsoidal to cap type is at an Eötvos number of 40. The drift flux V_{gj}^+ were determined by Ishii and Chawla (1979) for ellipsoidal type (eq. 3.17) and Kataoka and Ishii (1987) for the cap type (eq. 3.18 and eq. 3.19), respectively

$$V_{gj}^+ = \sqrt{2}(1-\alpha)^{1.75} \qquad \text{(ellipsoidal)} \tag{3.17}$$

$$V_{gj}^+ = 0.0019\left(D_H^*\right)^{0.809}\left(\frac{\rho_g}{\rho_f}\right)^{-0.157}\left(N_{\mu f}\right)^{-0.562} \qquad \text{(cap)} \tag{3.18}$$

for $D_H^* < 30$ and $N_{\mu f} < 2.2 \times 10^{-3}$

$$V_{gj}^+ = 0.030\left(\frac{\rho_g}{\rho_f}\right)^{-0.157}\left(N_{\mu f}\right)^{-0.562} \qquad \text{(cap)} \tag{3.19}$$

for $D_H^* > 30$ and $N_{\mu f} < 2.2 \times 10^{-3}$

D_H is the hydraulic diameter and $N_{\mu f}$ the viscosity number determined by

$$N_{\mu f} = \frac{\mu_f}{\left(\rho_f \sigma \left(\frac{\sigma}{g\Delta\rho}\right)^{0.5}\right)^{0.5}} \tag{3.20}$$

These relations have been experimentally verified for lead-bismuth and nitrogen mixture by Suzuki *et al.*, (2003).

The friction pressure gradient for the two-phase flow is expressed in the same form as in single-phase flow. The friction coefficients for single-phase are corrected with friction multipliers to account for the effect of a two-phase mixture. The evaluation of the pressure drop in the riser is accomplished using the Friedel model (1979), which is one of the most accurate two-phase pressure drop correlations and was obtained by optimizing the equation for the friction multiplier Φ_{lo} with large amount of experimental data. The model is valid for vertical upward flow and for horizontal flow and preferred for fluids with operating conditions that give a $(\mu_f/\mu_g) < 1000$ (Collier and Thome, 1998). The correlation has the following form

$$\Phi_{fo}^2 = A_1 + \frac{3.24 A_2 A_3}{Fr^{0.045} We^{0.035}} \tag{3.21}$$

where

$$A_1 = (1-x)^2 + x^2 \left(\frac{\rho_f f_{go}}{\rho_g f_{fo}} \right) \tag{3.22}$$

$$A_2 = x^{0.78} (1-x)^{0.024} \tag{3.23}$$

$$A_3 = \left(\frac{\rho_f}{\rho_g} \right)^{0.91} \left(\frac{\mu_g}{\mu_f} \right)^{0.19} \left(1 - \frac{\mu_g}{\mu_f} \right)^{0.7} \tag{3.24}$$

f_{go} and f_{fo} are the friction factors defined by equations for single-phase flow, μ_f the liquid viscosity, μ_g the gas viscosity, x the mass quality parameter, Fr the Froude number and *We* the Weber number.

3.5 Validation of Thermalhydraulics Models

In the 1D thermalhydraulics model equations are solved for energy and momentum using a finite volume method. The buoyancy terms of the momentum equation are calculated by integration of thermal gradients and mixture density difference along the loop. The flow rate and temperature field are solved by time stepping using the Euler implicit method.

The three submodels were coupled in a C++ program and the validation is performed by comparing the steady state results against the values provided in the conceptual design of an actinide burner cooled by SINGLE PHASE Pb-Bi natural circulation proposed by INL and MIT (Davis *et al.*, 2002). For the thermalhydraulics mesh, the total number of cells used was 170 (50 for the core, 50 for the heat exchanger, 10 for each section of lower plenum, upper plenum and the inlet of the core and 20 for the riser and downcomer each). For the fuel pin model, the total number of cells was 30, divided in 20 for the fuel and 10 for the cladding. The results of some design parameters for a 1300 MW_{th} and 650 MW_{th} core power are presented in table 3.1. The values obtained by our model show good agreement with the values reported by Davis *et al.* (2002).

Table 3.1: Comparison of results from the 1D model and the design values reported by Davis *et al.* (2002) for a single phase Pb-Bi cooled reactor

Parameter		1300 MWth		650 MWth	
		Davis *et al.*	This work	Davis *et al.*	This work
Tcore outlet	[K]	882	886	791	804
Tclad	[K]	938	926	876	875
Tfuel center	[K]	1088	1076	n.a.	951
mass flow rate	[kg/s]	23335	23380	17409	16816

The TWO-PHASE natural circulation model was compared with experimental results obtained in a LBE-Nitrogen gas loop at the Central Research Institute for Electrical Power Industry of Japan (Nishi *et al.*, 2003; Ceballos *et al.*, 2005). The test section is a closed loop formed by the connection of four cylindrical pipes which is filled with Pb-Bi. The inner diameter of the riser pipe can be changed from 155.2 mm to 106.3 mm and to 69.3 mm by inserting an additional pipe (that fits in the original riser pipe). The height of the free surface level of Pb-Bi is 1124 mm and the loop is kept at a constant temperature of 473 K, using an electrical heater with a thermal control unit. Figure 3.9 shows the scheme of the facility used for the experiments.

The experimental and numerical results are compared in figure 3.10. They show good agreement with the prediction of the LBE mass flow rate at low gas injection rate and for the larger pipe diameter. This is probably due to the bubbly flow regime obtained during the experiments. When the pipe diameter is reduced, the averaged void fraction is increased and the flow regime changes into transitional flow. Here the experimental results have a larger distortion and deviate from the predictions. The correlations implemented in the drift flux model are not appropriate under these conditions.

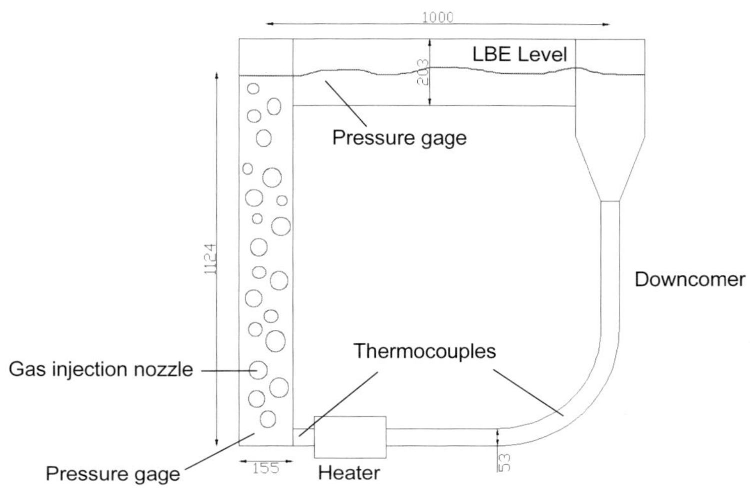

Figure 3.9 Description of the experimental facility at Criepi (Japan)

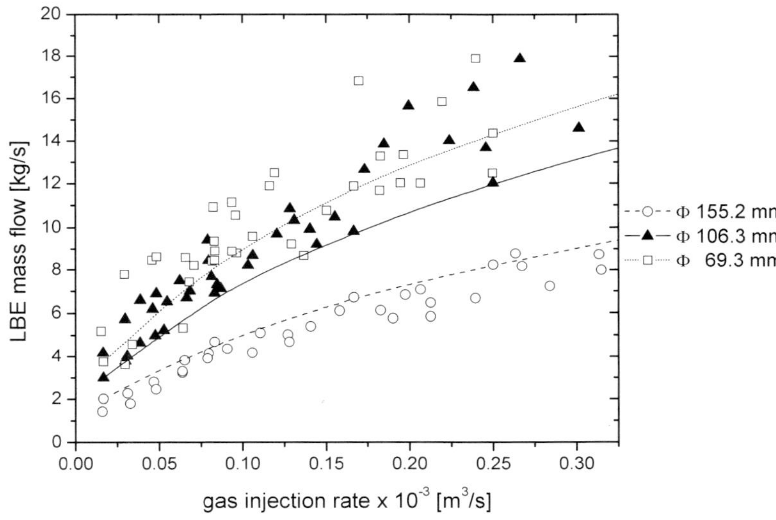

Figure 3.10 Comparison of experimental results and model predictions for the Criepi LBE loop

For the 2D thermalhydraulics model, several benchmarking exercises were carried out to ensure that the code provides the optimal solution in terms of accuracy and resources consumption. Some exercises aimed to optimize the number of zones for the radial and axial description of power distribution in the core, also the study of the time step size on the solution since the time scales of neutronics, fuel pin heat transfer and thermalhydraulics of the reactor pool were largely different.

The benchmarking of the code was based on the reactor conceptual design proposed by Rubbia *et al.*, (1995). The steady state values in this report were used as reference for the CFD results. Table 3.2 presents a comparison of the 1D solution, the 2D solution and the design values. For the 2D case, the temperature is averaged with respect to the mass flow rate at the same cross section high. The table indicates that the mass flow have a very good agreement as could be expected, considering that the main pressure drops are given by the reactor core and heat exchanger. These were implemented using the same friction correlations in both calculations. For the temperature, the values presented by the 1D model correspond well with the averaged values of the 2D case. This is also expected as the energy balance should be fulfilled. However, the spatial information provided with the 2D model has important added value if a detailed assessment of the fuel and cladding temperature is required, as well as the prediction of different phenomena regarding to the hydrodynamics of the reactor pool.

Table 3.2: Comparison of results of the 2D thermalhydraulics model against the 1D model and the design values reported by Rubbia *et al.* (1995)

CASE STUDY	Mass flow rate [kg/s]	Average temperature [K]	Maximum temperature [K]
2D – flat radial shape & axial uniform	51645.1	883.9	888.4
2D - cosine radial shape 6 x 4 cosine axial	51182.6	882.1	919.0
1 D model	51870. 9	871.1	871.1
CERN design values (Rubbia *et al*)	50000.0	873.1	873.1

3.6 Concluding Remarks

The purpose of this chapter was to present the physical models used for the safety analysis of the ADSR. The model combines three different phenomena to study the dynamic response of the reactor: neutron kinetics, fuel pin heat transfer and thermalhydraulics. The modeling strategy has approached the problem both with a simplified and a detailed version of the core. The benchmarking stage has compared modeling results with analytical solutions and/or with another computer codes. Enough evidence of the correct implementation has been provided. However, there is still further need for benchmarking exercises with thermalhydraulics experimental data, especially for heat transfer and buoyancy flow phenomena.

"We cannot be both the world's leading champion of peace and the world's leading supplier of weapons."
Jimmy Carter Jr., Peace Nobel prize 2002

4 Steady State Analysis of the ADSR

The inherent safety capability of natural circulation makes reactor design more reliable. Additionally, the construction and operation of a nuclear power plant with natural circulation in the primary cooling circuit is an interesting alternative for nuclear plant designers, due to the lower operational and investment costs obtained by simplifying systems and controls. This chapter deals with the application of natural circulation in the primary cooling circuit of an ADSR.

4.1 Introduction

The dependency on pumps and back-up systems to withstand the possibility of a loss of flow accident, leave the safe operation of the reactor to rely on external systems, which are expensive and require maintenance. Alternatively, physical phenomena such as natural convection could reduce or even eliminate this dependency, improving the economy and safety of a particular design.

In principle, natural circulation could assist the safety features of the ADSR (Rubbia *et al.*, 1995). But when natural circulation is implemented at commercial scale, the high core power causes a large temperature increase in the coolant. This is beneficial for buoyancy forces, but adverse for thermal stress. Since the material temperature limits bound the reactor power, the economical competitiveness is sacrificed. To overcome this problem, a higher coolant flow rate is required in order to remove more heat. This is achieved by increasing the vertical distance separating the heat exchangers and the core. Unfortunately, economical and technical issues such as manufacturing a large vessel onsite and plant size become other limiting factors. As a result, new alternatives for improving heat removal performance are being considered. To further enhance the liquid metal flow, injection of inert gases in the riser has been proposed (Ansaldo, 2001) and/or the design of a direct contact heat exchanger to minimize friction losses (Buongiorno *et al.*, 2001).

It is expected that the flow-enhanced reactor offer some of the advantages of the natural circulation single-phase reactor (Fomichenko, 2004). The injection of gas in the riser of the reactor increases the flow circulation by adding a buoyancy force to the fluid. The gas lift pump is more reliable since requires no moving components, it should hold an storage tank as well as a compressor, which supplies the required pressure for injection at the bottom of the riser. The storage capacity should sustain the flow rate for a few minutes during certain transient events.

Current research programs in liquid metal flow enhancement using gas injection have estimated the effect of the gas injection enhanced circulation in air-water system (Ambrosini *et al.*, 2005) and in a nitrogen-lead bismuth system (Nishi *et al.*, 2003). These results show the great benefit of the two-phase system over the single-phase natural circulation, even with small injection rates. This concept has been implemented in the conceptual design for the RBEC-M reactor designed by the Russian organizations OKB Gidropress, RRC Kurchatov Institute and IPPE (Fomichenko, 2004) and the experimental XADS (Ansaldo, 2001). Its use eliminates the hydraulic pumps dependence for reactor operation and reduces the pumping cost by approximately 6% (Fomichenko, 2004).

This chapter explores the feasibility of the ADSR cooled by buoyancy driven liquid metal flow. The approach takes as baseline the design from MIT and INL for a single-phase natural circulation lead-bismuth actinide burner (Davis *et al.*, 2002). In this reactor, the core power is restricted to maintain cladding temperature below established limits. The chapter consists of three sections. The first section presents the physical description of natural circulation liquid metal cooled reactors. The second discusses the implementation of the buoyancy enhancement with gas injection in a natural circulation liquid metal cooled ADSR. Finally, the third section looks at hydrodynamic effects that restrain the maximum operating power range.

4.2 Physics of Single-Phase Flow

The reactor primary cooling circuit is represented as a simple loop described in figure 4.1. When the coolant passes through the core and heat exchanger its temperature changes causing a density change. The density difference between the hot and cold legs derives a buoyancy force driving the circulation of flow in the loop. An steady flow rate is obtained when there is equilibrium between the buoyancy and the friction losses.

The flow rate is described by

$$\begin{gathered}\frac{dw}{dt}\sum_{n=1}^{N}\frac{l_n}{A_n}+\oint\rho g\beta\left(T-T_{ref}\right)dz+\\ \frac{w^2}{2\rho}\left(\sum_n\frac{K_n}{A_n^2}\right)+\frac{w^2}{2\rho}\left(\sum_n f_n\frac{l_n}{D_n}\frac{1}{A_n^2}\right)=0\end{gathered} \tag{4.1}$$

The buoyancy term of the momentum equation contains a temperature difference, which is determined by the energy equation in every zone, as it was presented in section 3.4.

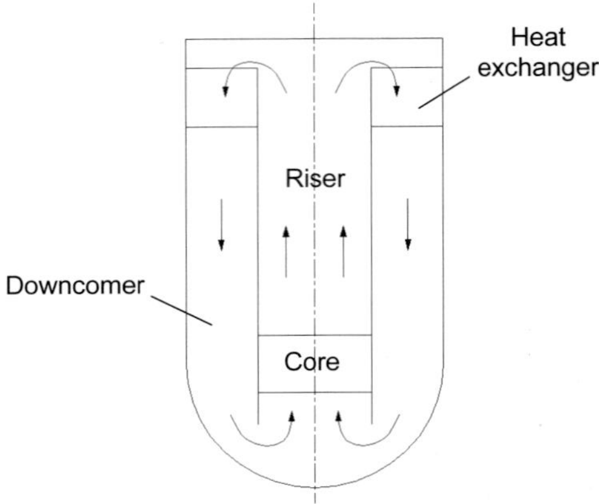

Figure 4.1 Description of the flow in a reactor pool

If the riser and downcomer are considered adiabatic, the energy balance over the core reads

$$C_p\left(\rho A\frac{\partial T}{\partial t}+w\frac{\partial T}{\partial z}\right)=q \tag{4.2}$$

Similarly, the energy balance over the heat exchanger reads

$$C_p\left(\rho A\frac{\partial T}{\partial t}+w\frac{\partial T}{\partial z}\right)=-h(T-T_w)s_{hx} \tag{4.3}$$

Non-Dimensional Analysis

Non-dimensional analysis is useful to study the fundamental physics of liquid metal single-phase natural circulation reactors. It identifies the physical properties and relevant parameters dominating the nature of the problem. The non-dimensional analysis allows understanding how the combined effect of fluid physical properties, characteristic lengths and temperature gradients limit the buoyancy flow. The selection of the fluid with best buoyancy characteristics can be an application of these results, however, it is unpractical due to the limited choice of liquid metals (Ceballos *et al.*, 2004). Nevertheless, the method can be further used to optimize the design.

To transform the equations 4.1 to 4.3 in non-dimensional form, one has to introduce the following non-dimensional parameters:

$$w_{ref} = \rho u_{ref} A_{ref} \qquad u_{ref} = \sqrt{g\beta\Delta TL} \qquad \Delta T = \frac{q}{hs}$$

$$t^* = \frac{tu_{ref}}{L} \qquad z^* = \frac{z}{L} \qquad T^* = \frac{T - T_w}{\Delta T}$$

$$w^* = \frac{w}{w_{ref}} \qquad A^* = \frac{A}{A_{ref}}$$

where u_{ref} is a reference velocity, w_{ref} a reference mass flow, L a reference length and A_{ref} a reference area. The reference length is chosen as the vessel height and the reference area is the cross sectional flow area of the core. Their substitution into equations 4.1, 4.2 and 4.3, leads to the non-dimensional equations 4.4, 4.5 and 4.6

$$\frac{dw^*}{dt^*}\sum_{n=1}^{N}\frac{l_n^*}{A_n^*} + \oint T^* dz^* + K_T \frac{w^{*2}}{2} = 0 \tag{4.4}$$

$$A^* \frac{\partial T^*}{dt^*} + w^* \frac{\partial T^*}{\partial z^*} = St_m \qquad \text{(core)} \tag{4.5}$$

$$A^* \frac{\partial T^*}{dt^*} + w^* \frac{\partial T^*}{\partial z^*} = St_m T^* \qquad \text{(heat exchanger)} \tag{4.6}$$

where T^*, w^*, z^*, A^* and t^* are the non-dimensional variables. The dimensional constants are grouped into the following parameters:

$$St_m = \frac{hLs}{w_{ref} C_p} \tag{4.7}$$

$$K_T = K_n + K_\mu \left(\frac{A_{ref}}{A_i}\right)^2 \tag{4.8}$$

where the Stanton number St_m is the ratio of heat transferred into a fluid and the thermal capacity of the fluid. For the loop case, it relates the systems capacity to dissipate the energy transported by the coolant. K_T is the total pressure loss factor of the system.

The solution of the steady-state problem can be obtained from substitution of the temperature for the core and heat exchangers (equations 4.5 and 4.6) inside the momentum equation (equation 4.4). Steady-state results are obtained by time-stepping the time dependent equations until the values no longer change to the specified value. The temperature profile obtained for the core is linear function of the core height, while for the heat exchanger is an exponential function. Similarly, the non-dimensional core outlet temperature and mass flow were calculated for different St_m and K_T numbers, the results are shown in figures 4.2 and 4.3.

The figures display that a decrease of pressure drop (lower K) causes an increase of mass flow rate and as a consequence a reduction of outlet temperature of the core. The temperature limit plotted in figure 4.2 corresponds to a given limit that ensures safe operation. Additional boundaries for figures 4.2 and 4.3 originate from technical and economical limitations. For liquid metal reactors, the coolant density and vessel height strongly determine the economics of the design. In particular, the building size, vessel volume and vessel wall thickness.

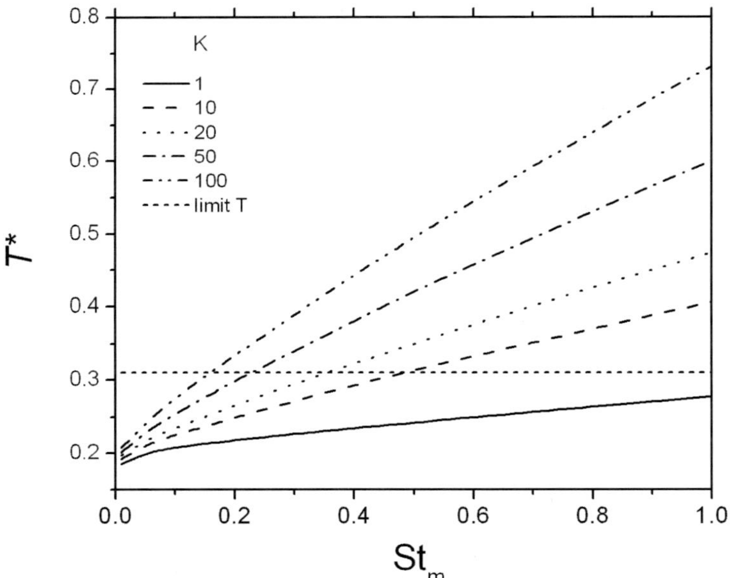

Figure 4.2 Stanton and friction numbers as function of the non-dimensional temperature in natural circulation flow

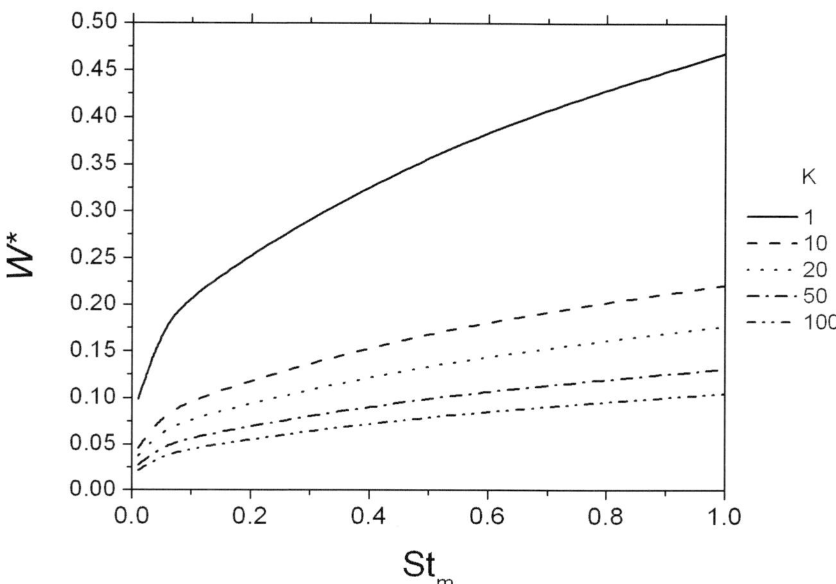

Figure 4.3 Stanton and friction numbers as function of the non-dimensional mass flow rate in natural circulation flow

The relevant fluid properties for the non-dimensional parameters are listed in table 4.1. The evaluation of St_m for lead and lead-bismuth at the same conditions shows a similar result since the properties are very similar. However, the temperature increase in the core is function of the St_m/w^* ratio (this can be obtained from the steady state energy equation). Hence, the temperature increase is influenced by the Reynolds number, which appears in the K_T factor. As a consequence, Lead presents a lower flow rate and therefore higher core outlet temperature.

Table 4.1 Coolant thermo-physical properties

Properties		PbBi	Pb	Na
β x10^{-3}	[1/K]	0.134	0.112	0.293
C_p	[J/KgK]	146	147.3	1263.9
ρ	[kg/m^3]	10080	10460	832
μ	[x10^{-3} kg/ms]	1.29	1.843	0.242

The limited choice of coolants for liquid metal cooling makes uninteresting to compare the performance of other liquid metals in natural circulation. But still, it is desirable to favor flow rate circulation to increase the core power capacity. The other alternative is the optimization of the geometrical parameters and flow conditions. In this regard, Davis *et al.*, (2002) have advanced remarkably looking at several aspects of the design choices. They have studied the core power limits of a natural circulation single-phase lead-bismuth actinide burner. Their conceptual design has expanded the core power, keeping the cladding temperature below established limits for steady state and loss of feed water transient.

4.3 Flow Enhancement Using a Gas Lift Pump

An alternative solution to the single-phase natural circulation is the gas lift pump. This technique is used in mixing reactors, mining technology and wastewater treatment plants (Bai and Thomas, 2001). The reliability of this pump is quite high since there are no moving parts, however, the economy of the system has to be balanced with regards to benefits of its application. For the ADSR, the gas could also be used to carry the inhibitors to control metal corrosion in the reactor vessel. The injection of gas in the riser of the reactor increases the flow rate by adding a buoyancy force to the fluid, thereby the heat removal capacity is increased.

The implementation of the gas lift pump needs to consider the injection point, hydrostatic pressure effect (vertical height in the pool), the selection of a suitable gas for enhancing purposes and the amount of gas to be injected. With regards to the injection point, the gas should not introduce any negative changes in reactor neutron balance, therefore the injection point is located above the reactor core. The gas selected is Argon due to its low chemical activity that avoids metal corrosion, low activation and radiotoxicity decay, good buoyancy enhancement capability and low cost. Regarding the amount of gas, some practical aspects are: *i)* the flow pattern required for maintaining a homogeneous and stable flow *ii)* limited LBE speed to minimize corrosion and erosion of structures and *iii)* injection costs.

The two-phase flow pattern in the riser should be kept in dispersed bubbly flow ensuring homogeneity in the mixture and avoiding possible flow instabilities and vibrations. It is possible that a slug flow regime brings strong fluctuations of the liquid flow rate and pressure. From fluid mechanics, the bubbly flow pattern corresponds to low void fractions. The bubble relative velocity is a competition between the forces acting on the bubble, which are the liquid drag, the buoyancy, and the surface tension. The bubble velocity increases with the equivalent bubble diameter. The bubble size is independent of the injection hole and gas composition (Bai and Thomas, 2001). At equal void fractions, smaller bubbles cause larger gas-lift efficiency as small bubbles have a low-rise velocity and a more evenly distributed gas concentration across the pipe. Moreover, with small bubbles the transition from bubbly to slug flow is postponed to larger values of injected gas (Guet, 2004) but, as the bubble rises up in the pool, the void fraction increases due to the hydrostatic pressure decrease.

In order to see the enhancement capabilities of the gas, a parametric analysis over Davis' conceptual design was performed. For this design, the maximum gas flow rate that produces bubbly flow patterns is 5.11 kg/s (when gas is injected at the bottom of the riser). This value was calculated with an average gas density for the full range of operating temperatures (from 473 K to 973 K). The maximum void fraction for a bubbly flow pattern was taken from the map flow of Simmer code. This map describes that void fractions larger than 0.3 involve transitional flow - the Simmer code has been experimentally validated for LBE-nitrogen mixture by Suzuki *et al.*, (2003). Figure 4.4 shows the results of maximum inner cladding temperature as a function of void fraction in the riser if the reactor power is increased. If the maximum temperature limit for cladding is 900 K, a reactor power of 1000 MW_{th} is pointed as the maximum core power allowed with overload capacity of 10%.

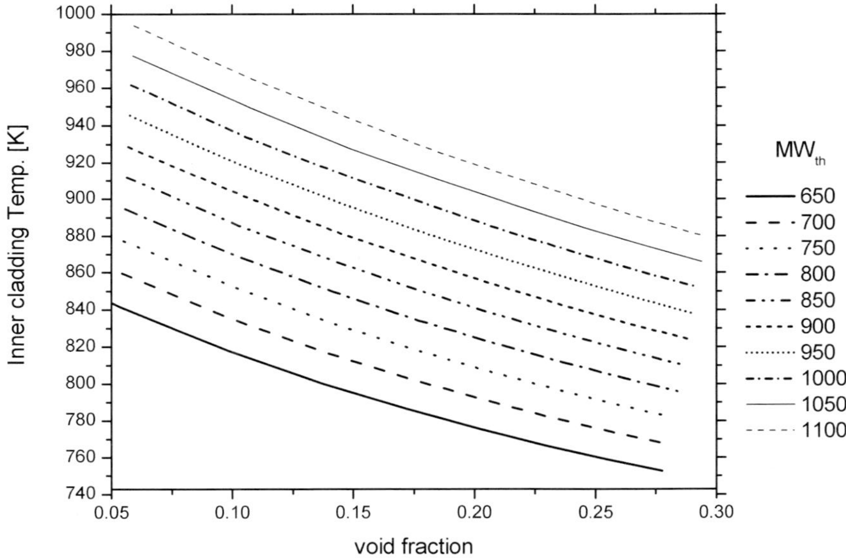

Figure 4.4 Inner cladding temperature as function of the void fraction

To further illustrate the benefits of buoyancy enhancement, figure 4.5 shows the effective temperature reduction across the core and the effective fraction of flow enhancement. The effective temperature reduction is defined as the ratio of core temperature difference from two-phase flow and single-phase flow. Since the power is increased, the effective comparison is weighted by the relative increase in core power. The effective flow enhancement is defined as the increased mass flow fraction of two-phase flow with respect to the single-phase flow.

The results show that for higher gas mass flow and constant reactor power, the effective temperature reduction increases. If the reactor power increases, the effective temperature reduction decreases. The most conservative case is the point with highest effective temperature reduction (45%) and same core power as the single-phase reactor (650 MW_{th}). The reactor with 1000 MW_{th} holds an effective temperature reduction of 30% and has economical advantages over the conservative case of 650 MW_{th}.

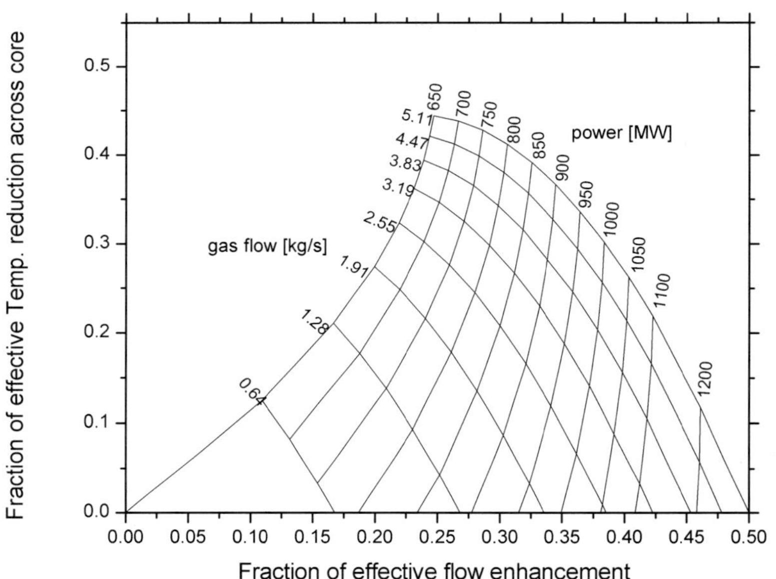

Figure 4.5 Fraction of effective flow enhancement as function of the fraction of effective temperature increase for different reactor power

Table 4.2 shows values of some design parameters for three different operating conditions: case 1 the single phase case with 650 MW_{th} core power (values from INL/MIT design), case 2 the enhanced flow concept with 650 MW_{th} core power and case 3 the enhanced concept with 1000 MW_{th} core power. The values show that gas injection allows for a significant increase in power while all temperatures remain within safe limits.

Table 4.2 Values of some design parameters in three different buoyancy driven concepts

Reactor Configuration		Single ph.	Two ph.	Two ph.
Power	[MWth]	650	650	1000
LBE Temp. core outlet	[K]	805	690	762
Max. cladding Temp	[K]	876	755	862
Fuel Temp. reference	[K]	951	830	978
LBE mass flow rate	[kg/s]	16816	29688	30835
gas mass flow rate	[kg/s]	0	5.11	5.11

The thermal behavior of the reactor as a consequence of the gas injection is characterized by the improvement of core cooling. The power can be increased and the reactor becomes more cost effective. In the same way, the margin to maximum temperature limits is larger because the core outlet temperature is lowered. Additional to this, the smaller temperature gradient across the core induces smaller peaks of thermal stress during transients.

4.4 Hydrodynamic Considerations

Thermalhydraulic analysis of the reactor core has three main objectives: the prediction of average flow characteristics, the prediction of local velocities and temperatures, and the prediction flow distribution and temperature behavior (IAEA-TECDOC-1060, 1999). As far as we are concerned, the complexity of the structure formed by a combination of fuel pins arranged in bundles makes safety analysis a difficult task. Consequently, the process of evaluating detailed hydrodynamic and heat transfer features in channels such as spacers effects, turbulence and non-uniform flow distributions is ignored. Instead, we are interested in studying the impact of power shape on the predicted fuel pin temperature. For this purpose, a model specifying the space dependent power shape in the core and the flow pattern in the reactor pool was implemented. The details were previously described in section 3.4. According to the modeling approach, the core has been divided in 6 radial by 4 axial zones. The temperature of the fuel pin is calculated by using local temperature conditions of the coolant and local heat transfer coefficient. The average coolant temperature is calculated based on mass flow fraction through the zones. The values for average coolant temperature at the core outlet and total mass flow are presented in table 4.3 and compared against the results obtained in the one-dimensional model used for preliminary calculations of section 4.3. The values of mass flow and temperature show a good agreement with the overall heat balance in the core. Note that the total amount of power was reduced from 650 MW_{th} to 450 MW_{th} because the initial results exceeded the maximum limits.

Table 4.3 Comparison of results from the 1D and 2D-axisymmetrical models

Parameter		2D axisym. 450MW_{th}	1D model 650MW_{th}
LBE mass flow	[kg/s]	27313	29688
Coolant outlet Temp.	[K]	654	690

Table 4.4 shows the values of inner cladding temperatures for a given radial and axial power distribution shape in the core. The values are reported for 24 different points in the 2D core geometry. The core radius is divided in 6 zones and the axial core distance in 4 zones (see figure 5.13). The distribution shape was taken from Rubbia *et al.*, (1995) and corresponds to a k_{eff} = 0.95. The resulting values of cladding temperature display a large gradient from central to outer zones of the core. When comparing the temperature of the inner ring at the top surface (radial 1 – axial 4) versus the outer ring at the same axial level (radial 6 – axial 4); the temperature difference is more than 300 K. The axial gradient is significantly lower in the core outer regions due to the low power deposited.

Table 4.4 Inner cladding temperature [K] at different zones of the core

	Radial	1	2	3	4	5	6
Axial	1	663	641	619	600	580	546
	2	781	732	690	653	613	551
	3	868	804	746	696	641	556
	4	898	832	770	714	655	560

4.5 Concluding Remarks

The performance of natural circulation is subject to a balancing system of buoyant forces and pressure drops. The non-dimensional analysis has shown that the thermal expansion, heat capacity and fluid density combine in the Stanton number. Relating the system's capacity to dissipate heat. Meanwhile, the fluid viscosity appears in the friction number. The flow enhancement using a gas lift pump can improve significantly the safety and economy of the design. The modeling results show a promising option of 450MW_{th} power.

"If we hope to escape self-destruction, then nuclear weapons should have no place in our collective conscience and no role in our security"

Dr. Mohamed ElBaradei, Peace Nobel Prize, 2005

5 Transient Analysis of the ADSR

Transient analysis is part of a design methodology to determine the safety and reliability of a specific design. For the present work, it is a practical exercise to understand the fundamentals of operation and safety in a buoyancy driven cooled ADSR. The objective is to characterize the reactor response of different thermalhydraulic concepts and to assess their safety during normal operation and accidents. Simplified and detailed models for reactor thermalhydraulic are used to cope with this objective.

5.1 Introduction

The dynamic behavior of the ADSR has been shown to be different from that of critical systems (Schikorr, 2001). Transient events with beam shutdown are expected to be safe if decay heat removal systems work. For most transients, a lower k_{eff} is preferred, because it holds a larger safety margin incase of positive reactivity feedbacks. Although, for transients without beam shutdown (unprotected transients), a higher k_{eff} (close to unity) could be advantageous since the Doppler effect would reduce the power, alike critical systems. This specific case has been studied by Van Dam (1999). He has demonstrated that a loss of heat sink without beam shutdown is unfavorable compared to a critical system. Complete shutdown will never be achieved if the beam remains on and the structural integrity is risked. Similarly, the mitigating effects by using negative reactivity feedbacks are poor in comparison to critical reactors, as it has been foreseen in section 2.2, therefore, for the protection of the ADSR, it is very important to assure the beam shutdown and the careful monitoring of k_{eff} and Doppler coefficient during the fuel cycle.

Regarding thermalhydraulics, the buoyancy driven flows have been subject of study during several years for light water reactors. The aim is to conceive an advanced reactor, which does not require primary circulating pumps. The technology transfer from light water to liquid metal reactors has opened numerous questions about the feasibility and advantages attained for this proposal. In recent years, Rubbia *et al.*, (1995) and Ansaldo (2001) have published the conceptual design of an ADSR cooled by buoyancy driven flows. Naturally, they have assessed the safety of their correspondent design and provided first evidence about benefits gained when compared to forced circulation systems. The natural convection flow seems to control the heat removal and may avoid any meltdown of the core. Still, the auxiliary decay heat removal system limits the capability of total core heat removed. Further investigations have shown that the transient and accident behavior in ADSR reveals potential safety problems in the framework of severe core disruptive accidents and core meltdown (Maschek *et al.*, 2003).

The present work aims to identify the operational risks of a buoyancy-driven cooled ADSR. The first section of the chapter compares the behavior of three different concepts facilitating the characterization of the dynamic response of buoyancy-cooled systems. The comparison is based on the conceptual design proposed by Davis *et al.*, (2002) and the extended versions including a gas lift pump as discussed in table 4.2. The transients studied are caused either by variations of the proton beam of the accelerator, electrical systems failure or caused by a reactivity perturbation in the core. The subsequent section presents a detailed geometrical model. It includes the accurate geometrical representation of reactor pool and the local flow conditions in the core given a power distribution shape. In this way, the fuel pin temperature is calculated at different positions of the core. The final section of this chapter demands a comparison of 1D and 2D-axisymmetrical modeling approaches reviewing result similarities and differences.

5.2 ADSR Characteristic Transient Response

This section characterizes the reactor response of the three different concepts suggested in table 4.2. The selected scenarios are those of normal plant operation such as reactor start-up, beam trip, beam over power and some upset, emergency and faulted events like the loss of gas lift, the loss of heat sink, a reactivity insertion, the unprotected transient loss of gas lift and the unprotected loss of heat sink. For computational purposes the reactivity is considered as a relative measure of the effective multiplication factor with regards to the critical state, the subcriticality state is set to 0.95 following the recommendations proposed by Jovanovich (2004) and the values for reactivity feedbacks and neutron kinetic parameters are listed in appendix 1.

Reactor Start-up

The start-up simulation is important since the reactor must allow the slow expansion of materials to avoid thermal stress formation. Hence, for a start-up simulation, the gradient between inner and outer temperature of the fuel pin cladding was used to control the neutron source intensity. The simulation begins from a reactor shutdown condition in which the power level is at 2% of the original steady state full power (650MW_{th} and 1000 MW_{th}). The LBE mass flow rate for the two-phase flow reactors at shutdown conditions is very similar since decay heat power does not have a remarkable influence. The initial values are determined by the gas mass flow injected in the riser and the small buoyancy effect created by temperature gradients from the core decay heat. On the other hand, the mass flow rate of the single-phase reactor is given by the small temperature gradient only.

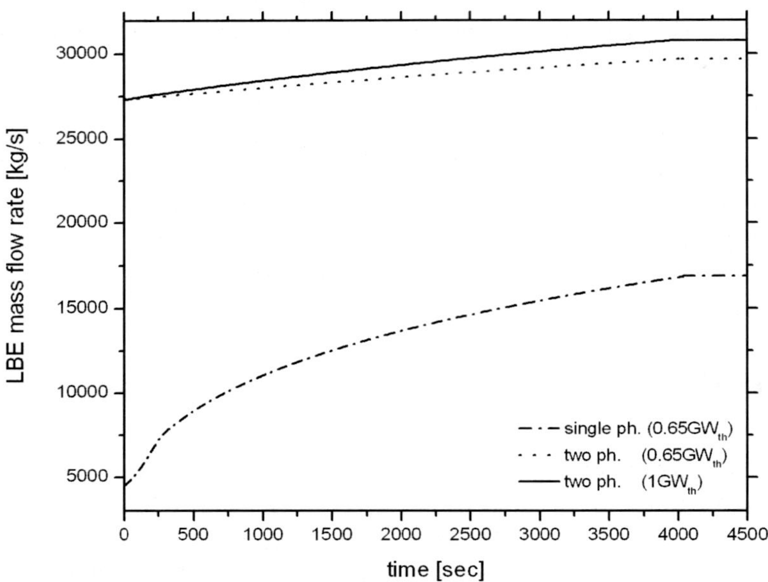

Figure 5.1 Mass flow rates during reactor start-up of three different designs: 650MW$_{th}$ single-phase, 650-1000 MW$_{th}$ two-phase

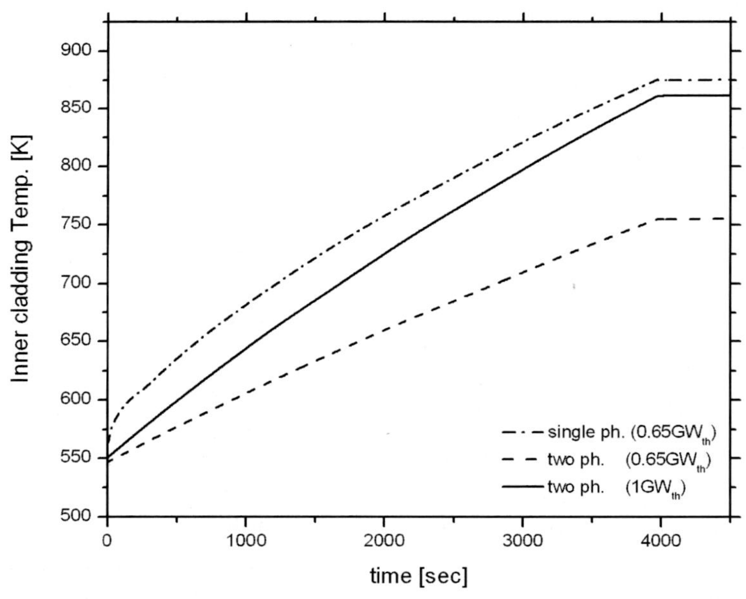

Figure 5.2 Cladding inner temperature during reactor start-up for the three different designs: 650MW$_{th}$ single-phase, 650-1000 MW$_{th}$ two-phase

The reactor power evolves from decay heat values to steady state full load. As the power increases, the equilibrium between friction and buoyancy is broken and the mass flow rate increases with the increase of the temperature gradient between the hot and cold leg. When full power is reached, the flow rate obtains a new steady state condition. For the two-phase configurations, there is a smoother transition to steady state since the flow is already close to it and has more inertia (see figure 5.1). Figure 5.2 shows that the increment on the inner cladding temperature during the start-up is smaller than 1 K per second, which is sufficiently slow to avoid thermal stress formation. Other structural components should follow a similar assessment.

Beam Trip

The beam trip is advantageous for rapid core shutdown. This transient is able to bring the ADSR in a few seconds from large power to decay heat levels. The initial condition for this transient is the steady operation at full power. At time zero, a sudden failure causes the immediate shutdown of the proton beam, in this manner, the neutron source is stopped and the power drops rapidly. A positive reactivity feedback is caused by the power drop and consequent fuel temperature drop, but the total reactivity remains well below zero ensuring a subcritical state.

For the single-phase reactor, the cold stream in the downcomer, enters the core but the temperature does not increase significantly because the power has decreased. When the hot coolant from the riser goes into the heat exchanger, it gets cooled. Then, the buoyancy by thermal gradients is diminished. The same effect occurs in the two-phase system, but the buoyancy induced by the gas liquid mixture remains. When the core outlet temperature drops, the gas density increases causing a decrease in the void fraction as well. A new stationary condition is encountered as it is shown in figure 5.3(top). Regarding the fuel pins, the power decrease causes a temperature reduction. The cladding temperature drops 200 K during the first 10 seconds of the transient. The cladding temperature drop is considered severe and its large value is a consequence of the large coolant temperature gradient between the core inlet and outlet. From the results, it is also clear that the time constants for heat transfer in the fuel pin and coolant flow are very different. The coolant mass flow rate is reduced slowly, and hence, the structural components under its flow path will not experience rapid thermal shocks such as those on the fuel pin. The cladding material might be subjected to large thermal peaking stress.

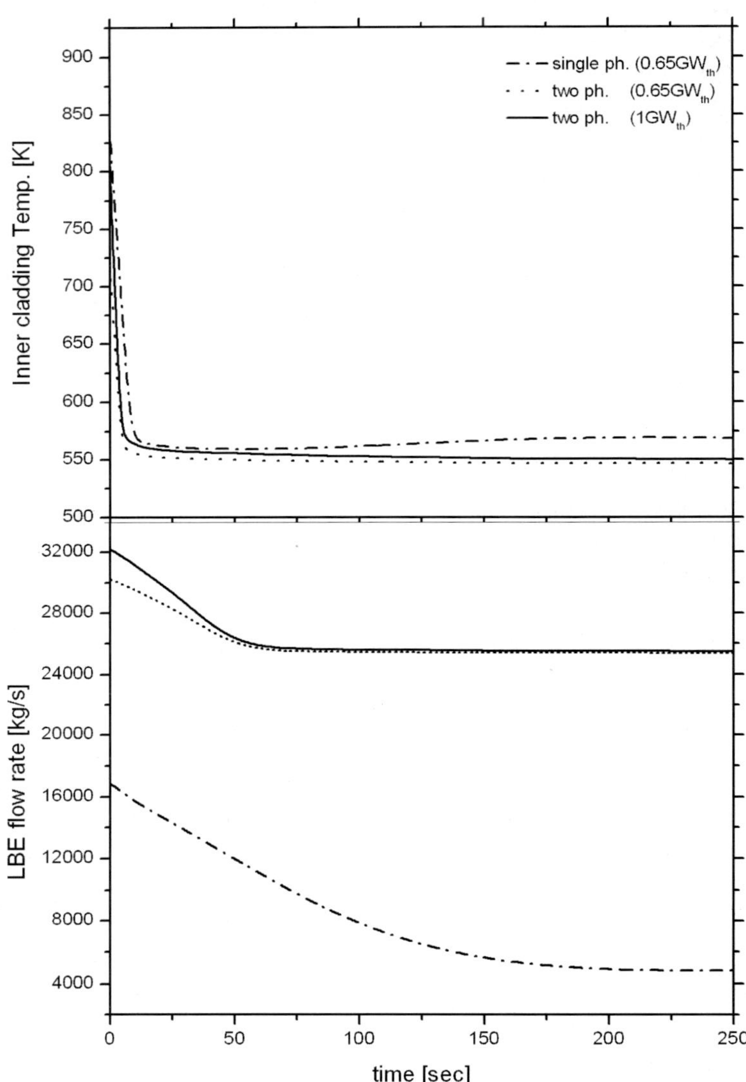

Figure 5.3 (top) Inner cladding temperature, (bottom) Mass flow rates, during a beam trip for the three different designs: 650MWth single-phase, 650-1000 MWth two-phase

Overpower

The overpower transient is a hypothetical transient that assumes a failure where the neutron source intensity is doubled for a period of 10 seconds, while the reactor was operating at steady full power. This transient simulation mainly attempts to identify the grace time before cladding temperature limits are exceeded. In this way, the margin for control systems or actuators to stop the proton beam could be defined.

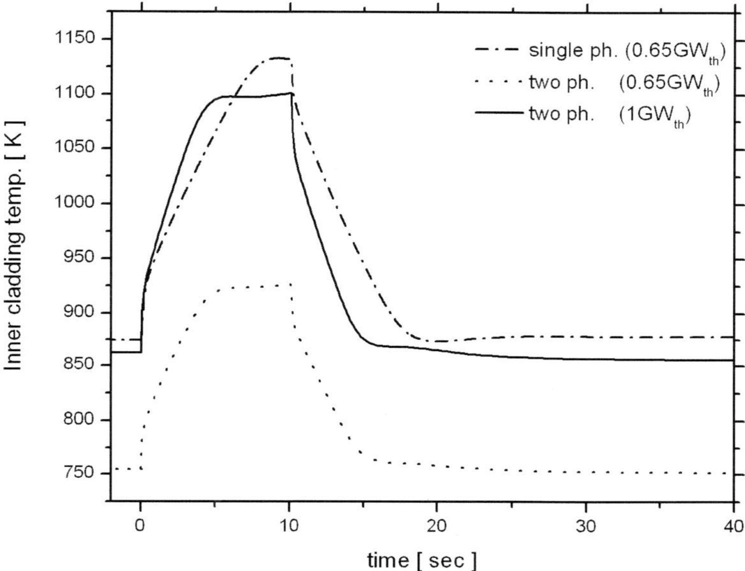

Figure 5.4 Inner cladding temperature during a beam overcurrent transient for the three different designs: 650MWth single-phase, 650-1000 MWth two-phase

The doubling of the neutron source for the ADSR causes a prompt response in power. After this, the negative reactivity feedback reduces the power increase rate towards a new asymptotic level. The effect of the negative feedback in the ADSR is moderated, since the reactor cannot be self-shutdown while the neutron source remains on. After 10 seconds of start, the magnitude of the neutron source returns to the original steady state value. A prompt power decrease is induced and is followed by a positive reactivity feedback. Figure 5.4 shows the temperature peak in the cladding caused by the overpower excursion. The peak temperature is 200 K above the initial value. The maximum temperature limits allowed for cladding materials are exceeded in the single phase and in the 1000 MW_{th} two-phase reactor. The heat discharged during the overpower excursion does not affect the LBE flow rate significantly. The circulation time in half loop is longer than the overpower excursion, thereby, only a small volume of coolant flows upward with higher temperature. For the two phase reactor, the hot liquid entering the riser, decreases the gas density and increases the void fraction, which enhances the flow slightly more.

Loss of Gas Lift

The simulation of loss of gas lift transient is only possible for the enhanced natural circulation reactor. This transient is expected to occur during reactor lifetime with a low probability per year. The initial condition for the loss of flow transient consists of steady-state full power operation. Then, a sudden failure shuts the gas mass flow down and the flow enhancement disappears. To prevent any damage of the core, the control unit stops the proton beam and the reactor power is dropped to decay heat levels.

Figure 5.5 shows the sudden decrease of the LBE mass flow following the gas lift shutdown. After the buoyancy force from two-phase fluid mixture disappears and the power drops, the LBE mass flow is reduced until it finds a new stationary condition. During the transition, the inertial force and buoyancy of thermal gradients finds a new equilibrium with the single-phase system's pressure drop. The transient does not affect the power, reactivity, fuel and cladding temperatures significantly, since the core shutdown resembles a beam trip transient.

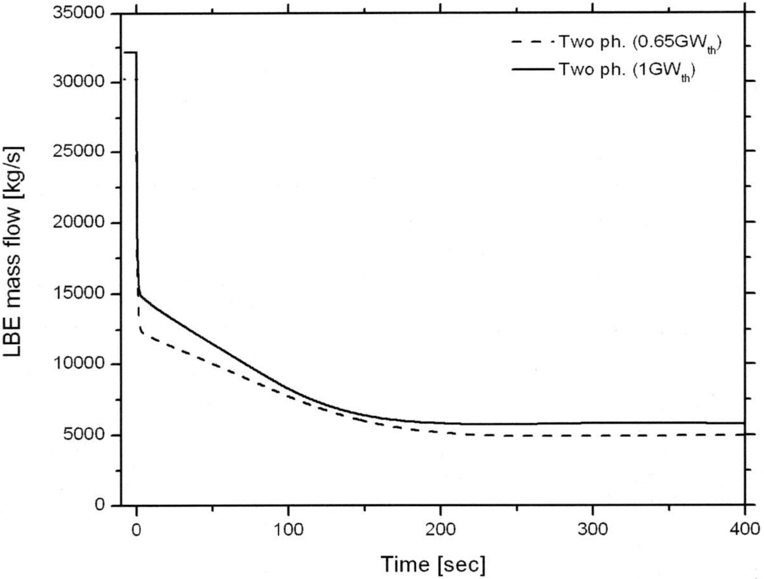

Figure 5.5 LBE mass flow rate during a loss of gas lift transient for the 650-1000 MWth concepts

Chapter 5

Loss of Heat Sink

The loss of heat sink transient studies what would happen if normal heat removal through the heat exchangers is not longer available. The initial condition for the simulation is the reactor steady state full power operation. Then, the heat exchanger feedwater pump fails and heat removal stops. The control unit orders the proton beam to stop and the power drops immediately. Heat removal through a reactor vessel auxiliary cooling systems (RVACS) is started in order to avoid core damage.

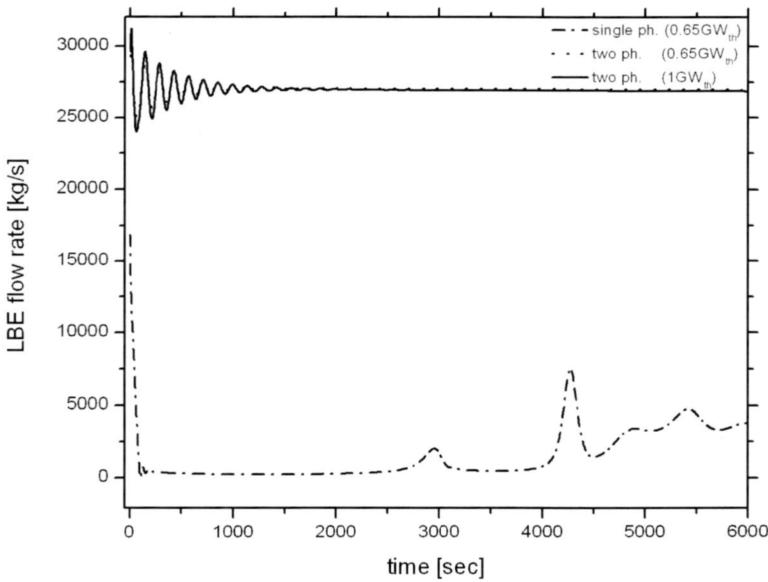

Figure 5.6 Coolant mass flow rate during a loss of heat sink transient for the three different designs: 650MWth single-phase, 650-1000 MWth two-phase

In the single-phase reactor, the hot coolant flowing inside the riser enters the heat exchanger and flows down hot in the downcomer. Similarly, the "colder" coolant that was inside the downcomer enters the core, but does not undergo a significant temperature increase. The buoyancy force decreases as the hot and cold stream go ahead inside downcomer and riser, respectively. In this regard, the direction of the buoyancy force is inverted against the inertial force. The LBE mass flow rate drops to a minimum as the hot stream has reached the bottom of the downcomer. Then, it flows again back into the core and riser, while the cold stream in the heat exchanger and downcomer. Now, the direction of the buoyancy is the same of the inertia, the flow gets a slight increase but friction forces manage to establish a new equilibrium. After this moment, the LBE temperature starts increasing and in the same way, the fuel and cladding temperatures do. The temperature increase builds up new temperature gradients and the buoyancy is recovered (see figure 5.6).

For the two-phase flow reactor, the coolant experiences similar temperature changes as in the single-phase reactor, but the gas mass flow rate holds sufficient buoyancy to keep a large percentage of the LBE mass flow circulation. Additionally, the LBE temperature changes affect the gas density in the riser and in this way, the void fraction quantity. The buoyancy given by two-phase flow oscillates until a thermal equilibrium is established at the core outlet temperature.

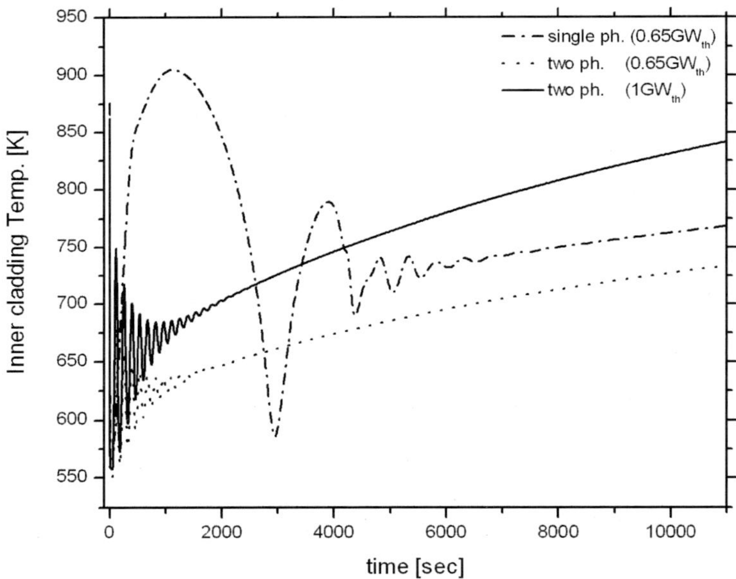

Figure 5.7 Inner cladding temperature during a loss of heat sink transient for the three different designs: 650MWth single-phase, 650-1000 MWth two-phase

Figure 5.7 shows the oscillatory pattern of the inner cladding temperature, which is caused by the oscillations of the LBE mass flow. The temperature peaks do not exceed the allowed temperature limits for cladding suggested in table 2.4; however, the oscillatory behavior could cause thermal fatigue by thermal cycling and failure of the fuel pins. The reactivity presents the same trend of the cladding temperature. The Doppler reactivity feedback from fuel temperature changes does not cause a large reactivity increase. The total reactivity is negative in all peaks assuring the subcritical state.

Reactivity Insertion

This transient gives an estimate of the grace time for the controls to shutdown the beam. Some possible sources of a positive reactivity insertion are: presence of voids in the core or the cladding removal. For the transient simulation, the initial conditions are the steady state full power reactor operation and then, a positive reactivity of 1 $ is inserted. The result of reactivity feedback calculations has shown that the amount of reactivity inserted and the selected subcriticality level do not imply criticality risks. After the insertion, the Doppler effect introduces a negative feedback that does not bring major effect over the power since the beam is still on. As a result, the increase of power is followed by an increase in fuel and cladding temperatures. Figure 5.8 shows that the inner cladding temperature limits are exceeded after some seconds of start in the single-phase and two-phase reactor of 1000 MW_{th}. The two-phase 650 MW_{th} has also crossed the limiting values, but its lower operating temperature offers a longer grace time.

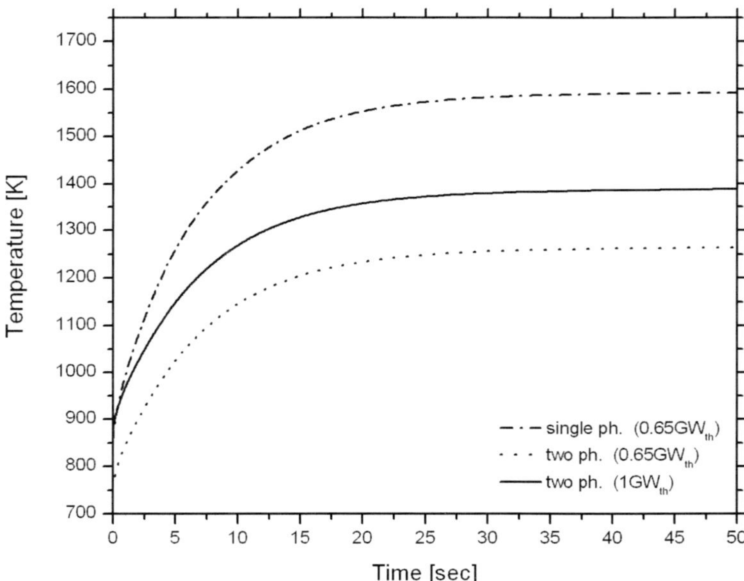

Figure 5.8 Inner cladding temperature during a reactivity insertion of 1 $ for 3 different designs: 650MWth single-phase, 1000 MWth two-phase

Unprotected Loss of Gas Lift

The initial condition for this transient is similar to the previously explained loss of gas lift transient. This time, the transient considers the failure to shutdown the neutron source. Under these conditions, the LBE mass flow rate is also reduced by the loss of buoyancy from the two-phase flow but the power continues heating up the coolant. Hence, fuel, cladding and coolant temperature increase. The fuel temperature increase is followed by a Doppler reactivity feedback, the total reactivity remains well below critical conditions.

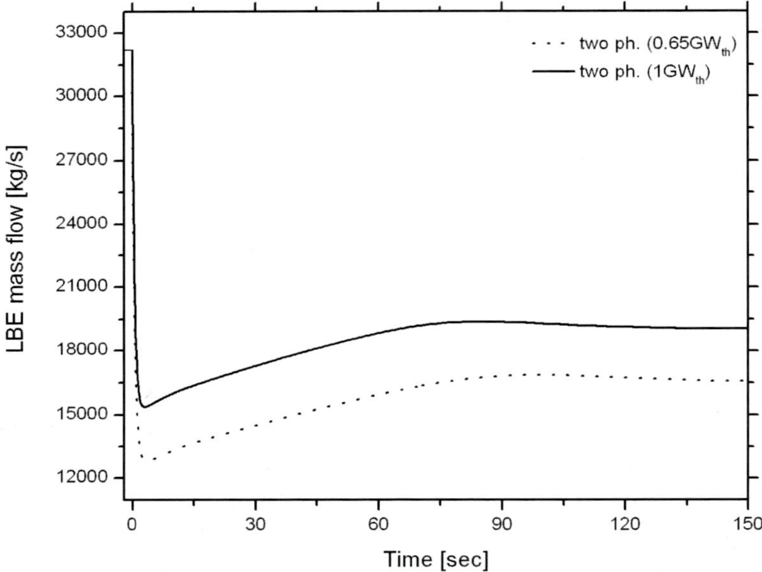

Figure 5.9 LBE mass flow rate during an unprotected loss of gas lift transient for the 650MWth and 1000MWth

Figure 5.9 shows that the LBE mass flow rate gets a new stationary condition after the gas lift is stopped. The new steady-state is given by the buoyancy from thermal gradients, which increases with the temperature increase in the core. In the same way, the cladding operating temperature shows a peak increase due to the gas lift loss. Then, as the LBE mass flow rate starts increasing and Doppler effect reducing power, the cladding temperature is reduced. A new steady value is found after the coolant has circulated once through the entire loop. The cladding temperature peak in the reactor of 1000 MWth has exceeded the maximum cladding temperature limit (see figure 5.10).

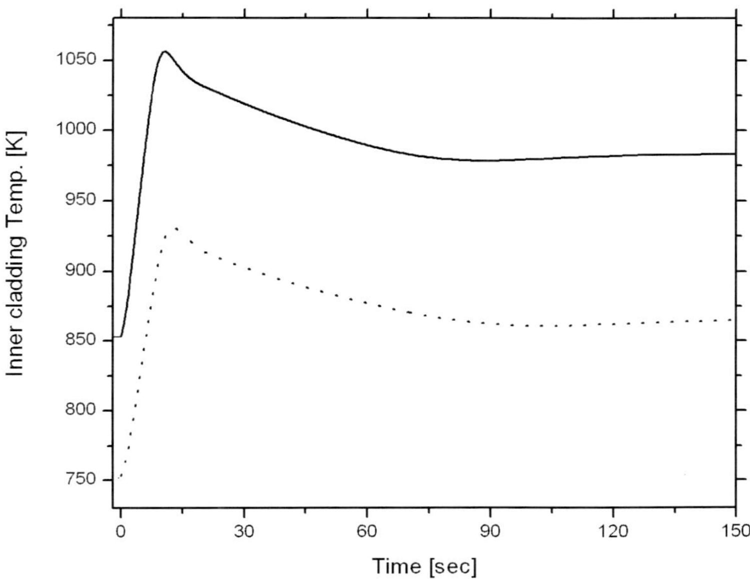

Figure 5.10 Inner cladding temperature, during an unprotected loss of gas lift transient for the 650MW$_{th}$ and 1000MW$_{th}$

Unprotected Loss of Heat Sink

The unprotected loss of heat sink transient involves the same events explained for the loss of heat sink transient but assumes that the proton beam is not shutdown. The reactor shows a reduction of buoyancy force because the hot stream flows downward in the heat exchanger and downcomer. As the LBE mass flow rate is reduced, the fuel temperature increases causing a negative Doppler reactivity feedback. The power is reduced, but the temperatures of fuel, cladding and coolant continue to increase because of the limited heat removal.

In the single-phase reactor, the "colder" stream from the downcomer enters the core and gets heated up. Two hot streams have filled the riser and downcomer of the reactor pool decreasing the LBE mass flow rate. Thus, when the hotter stream completes one circulation along the loop and enters the core, it is heated-up further and the buoyancy force and mass flow are recovered partially (see figure 5.11). For the two-phase reactors, the buoyancy from the gas liquid mixture is still sustained and then, the coolant takes less time to complete one circulation in the loop.

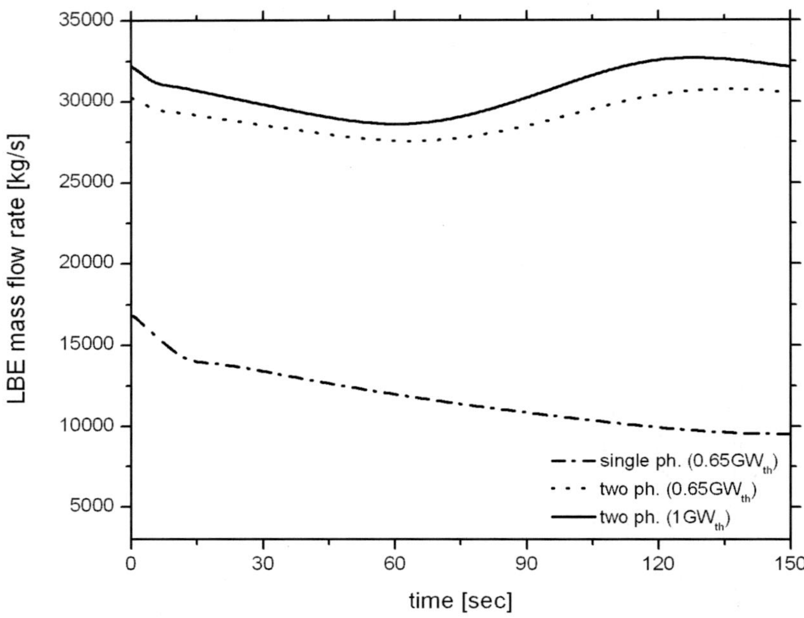

Figure 5.11 LBE mass flow rate during an unprotected loss of heat sink transient for the three different designs: 650MWth single-phase, 650-1000 MWth two-phase

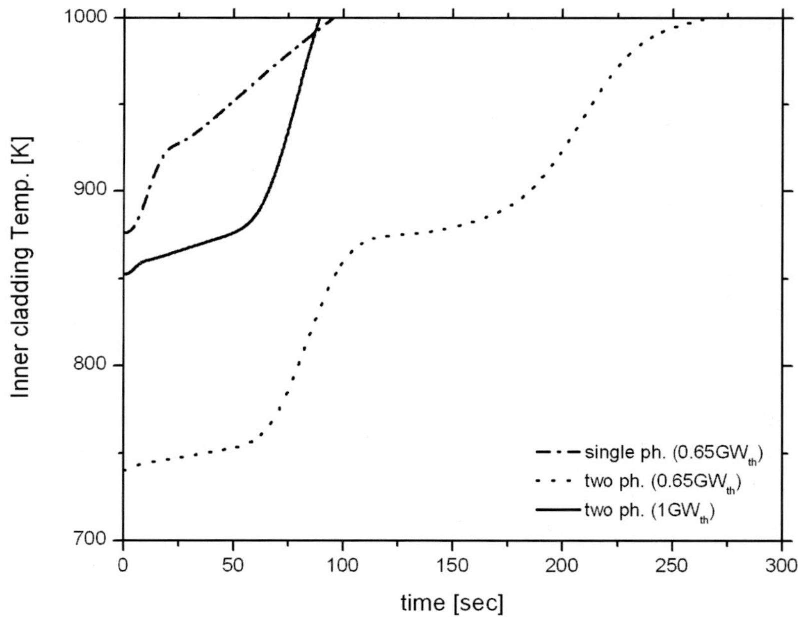

Figure 5.12 Inner cladding temperature during an unprotected loss of heat sink transient for the three different designs: 650MWth single-phase, 650-1000 MWth two-phase

Figure 5.12 shows the cladding temperature increase given by a higher coolant core inlet temperature and a lower mass flow rate of coolant. The increase is relatively fast, the maximum cladding limit is reached after the first minute of the transient for the single-phase reactor and for the enhanced circulation with 1000 MW_{th}. Additional actions should ensure the proton beam shutdown before this time.

5.3 Transient Analysis of the ADSR (Local Heat and Flow Conditions)

The previous analysis has investigated the transient response of different ADSR thermalhydraulic concepts. The results have shown that the two-phase enhanced natural circulation of 650 MW_{th} is safe for all transients except for the positive reactivity insertion of 1$ and the unprotected loss of heat sink. However, the grace time is enough to allow additional actions for beam shutdown. The safe operation of the 650 MW_{th} single-phase and the 1000 MW_{th} two-phase enhanced natural circulation reactors is questioned. The cladding temperature limits are reached during the following transients: overpower excursion, positive reactivity insertion of 1 $, unprotected loss of heat sink and unprotected loss of gas lift (only for the 1000 MW_{th}). Additionally, some of the calculated transients have shown that the cladding and structures could be subjected to high cycling thermal loads in all three designs. Following this, a detailed two-dimensional model integrating thermalhydraulic and neutronic calculation and capable of providing data for post processing structural thermal stress was developed.

In this section, the safety of the buoyancy-enhanced reactor is studied under selected events expected to occur during ADSR operation. To facilitate the comprehension of results from the 2D-axisymmetrical model, the core zones have been numbered with a two digit index, the first number corresponds to the axial position, from the core bottom to its top. The second index corresponds to the radial position, from the core centre to the outer side. Figure 5.13 displays the index distribution, i.e. TC11 corresponds to the cladding temperature of the bottom zone, inner ring of the core. TC46 corresponds to the cladding temperature of the upper zone and outer radial region.

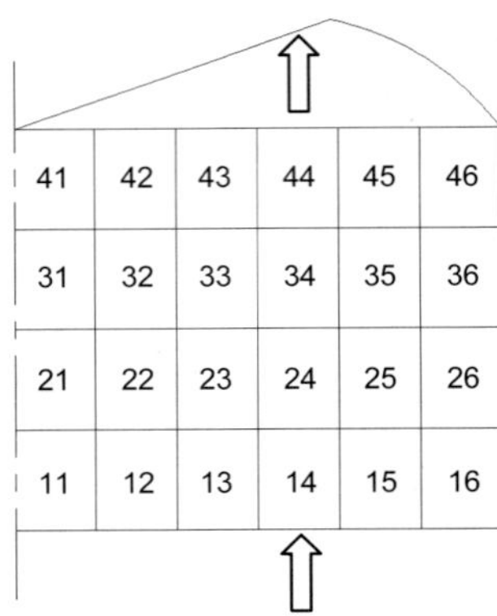

Figure 5.13 Index distribution to interpret local data results from the 2D-axisymmetrical model

Beam Trip

For the coming analysis, it is important to note that during the steady-state analysis in chapter 4, the 2D-axisymmetrical thermalhydraulic model determined that the inner-cladding temperature was under estimated severely by the one-dimensional model. For this reason, we were forced to reduce the core power from 650 to 450 MW_{th} in the following calculations. Hence, the initial condition of the beam trip transient are the steady full power operation at 450 MW_{th}. Then at *t=35 seconds*, the beam trips and the power drops promptly, as a consequence of the subcritical state inherent of the reactor core. The power falls below 20% from the initial steady value in less than 1 second. This response assures a complete shutdown. The Doppler effect produces a positive reactivity feedback rising k_{eff} from 0.95 in steady full power to 0.955 at decay heat conditions (figure 5.14).

When the fission process is stopped the coolant is still flowing around the fuel pin and thereby the fuel and cladding temperature drop considerably. Figure 5.15 shows the values of inner cladding temperature as a function of the axial position at the central zone of the core. The temperature decrease is very significant. The cladding wall undergoes a rapid temperature drop (in about 10 seconds) from maximum to minimum temperature values. The same effect occurs at the core radial zoning, but the outer zones present a moderate drop in comparison to the central regions. Figure 5.16 shows the temperature drop at different radial zones, the differences are a consequence of the power distribution profile in the core.

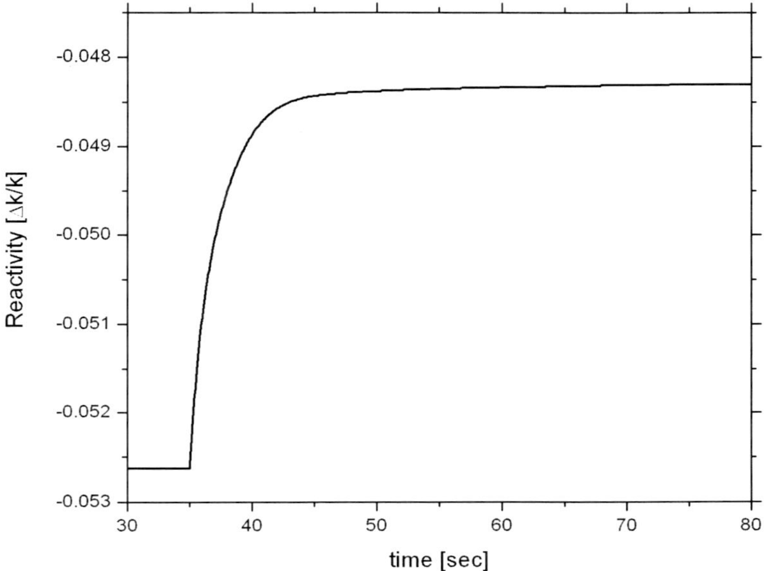

Figure 5.14 Reactivity during a beam trip in a 450MW$_{th}$ two-phase flow reactor

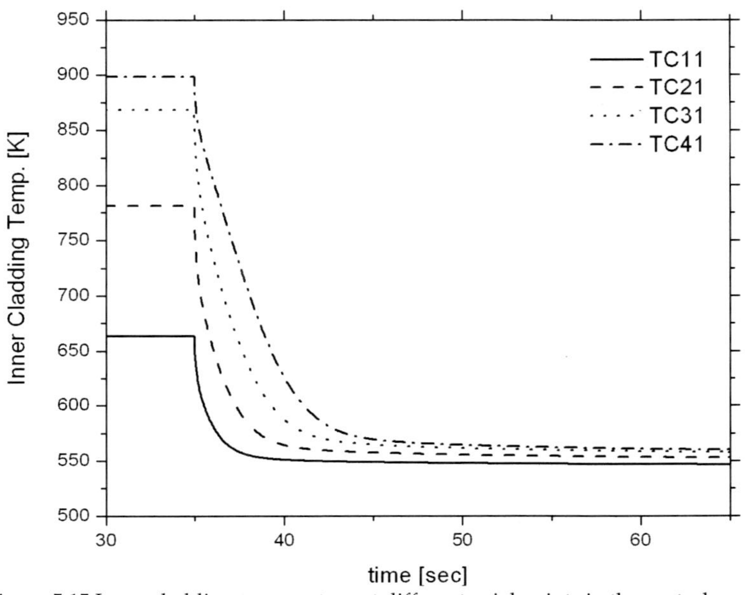

Figure 5.15 Inner cladding temperature at different axial points in the central zone of the core, during a beam trip in a 450MW$_{th}$ two-phase flow reactor

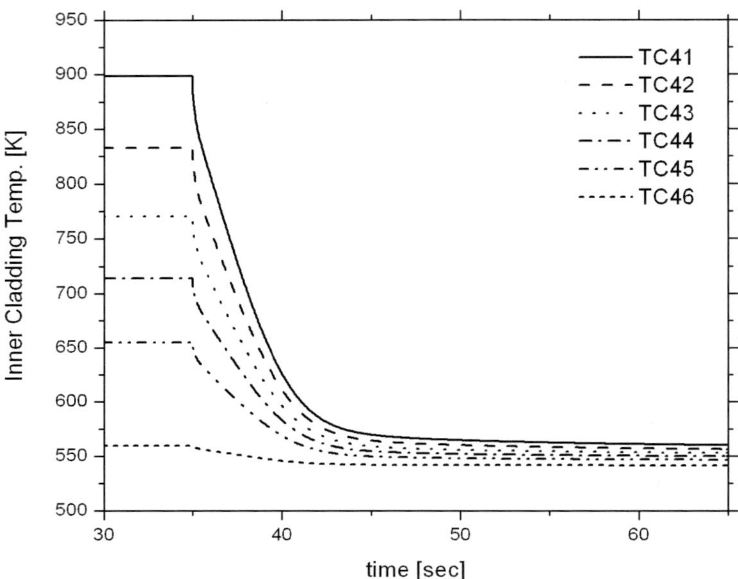

Figure 5.16 Inner cladding temperature at different radial points in the core upper zones, during a beam trip in a 450MW$_{th}$ two-phase flow reactor

Beam Interruptions

In particle accelerators, beam interruptions occur with large regularity and having different time length ranging from seconds to several minutes. They are consequence of the technical limitations in the present accelerator technology. For the ADSR, it is important to assess the behavior under different beam interruptions and their implications on safety and reliability. For this reason, we have included the analysis of different interruptions length.

The first interruption considers a time interval of 1 second from the beam trip to its return. Similarly, time intervals ranged from 15 seconds, 30 seconds, 1 minute and 5 minutes were arbitrarily chosen but considering that the number of interruptions increases as the time interval becomes smaller (Eriksson, 1998). Also, beam interruptions involving periods longer than 5 minutes were not considered since the expected results will not differ much from the 5 minutes case. Figures 5.17 shows the power response and reactivity feedback after the beam interruption of 1 second. At the start of the transient, the power falls below 20% of the initial power. During 1 second it has already arrived around 10% of the initial value. When the beam power is back, the power rises rapidly and the initial stationary conditions are re-established. The positive Doppler reactivity feedback consequence of the fuel temperature drop does not include any criticality risk. The time period needed to reach initial conditions followed the back power is not less than 10 seconds.

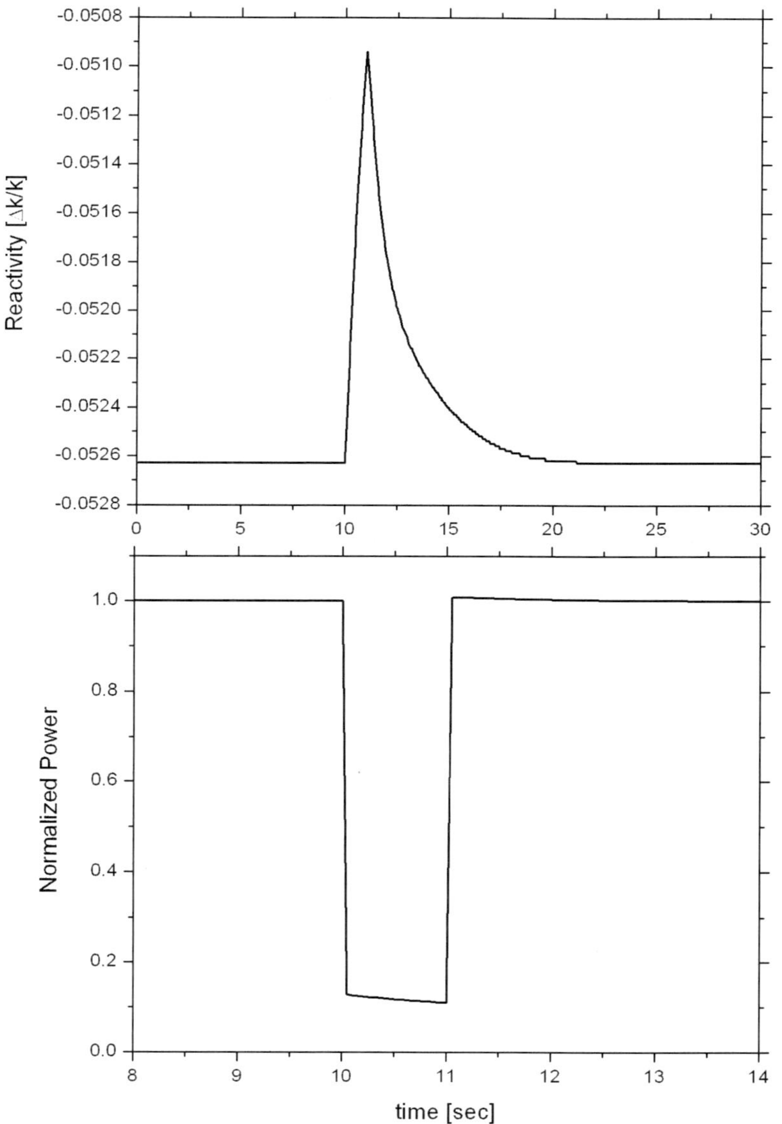

Figure 5.17 Reactivity (top) and Power (bottom) during a beam interruption of 1 second length in a $450MW_{th}$ two-phase flow reactor

Figures 5.18 show the effect caused on the cladding temperature during the beam interruption of 1 second. The figure displays large peak differences consequence of the power distribution profile. The temperature drop at the core central zones is deeper meanwhile, the effect at different axial locations present different pattern consequence of the change in coolant mass flow parameters.

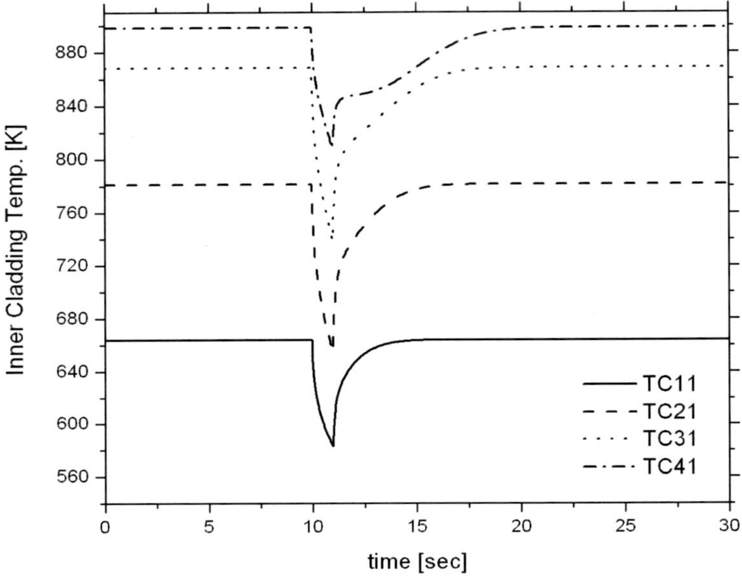

Figure 5.18 Inner cladding temperature function of axial position during a beam interruption of 1 second in a $450MW_{th}$ two-phase flow reactor

Figure 5.19 shows the cladding inner temperature in time for different radial and axial positions during 15-seconds and 1-minute beam interruptions. In contrast to figure 5.18, it indicates a larger temperature drop due to the longer time interval. One observes also that there is almost a vertical drop or jump following the beam trip or step, this event is a potential inducer of thermal stress and fatigue for the cladding. The results obtained from a beam interruption of 5 minutes are equivalent to the beam trip. The positive Doppler feedback demonstrates that the subcriticality is preserved and the return to normal operating conditions does not include long delays due to feedback effects. The coolant mass flow rate presents a small oscillation caused by void fraction change due to temperature fluctuations. The lower coolant temperature increases the gas density and hence, the void fraction reduces. The effect is temporal since a new thermal equilibrium is established rapidly (see figure 5.20).

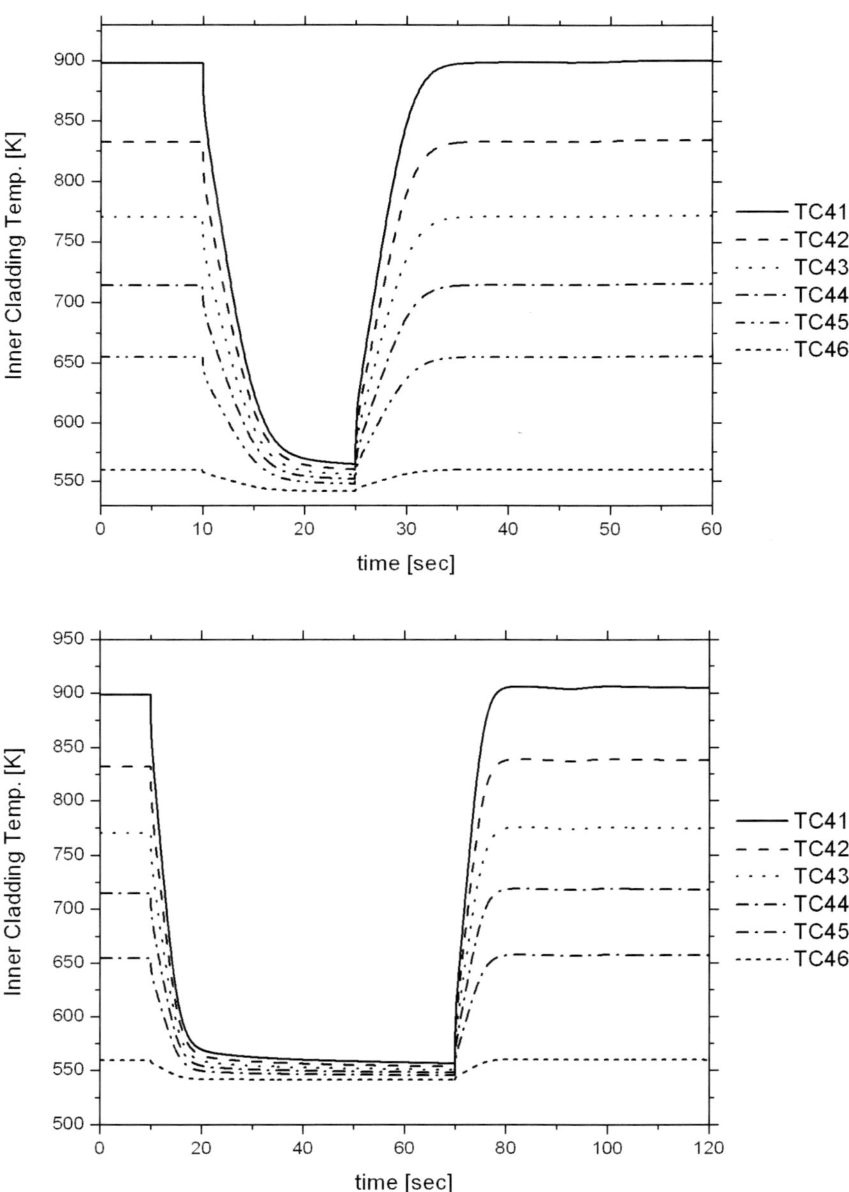

Figure 5.19 Inner cladding temperature during a beam interruption of 15 seconds (top) and 1 minute (bottom) in a 450MW$_{th}$ two-phase flow reactor

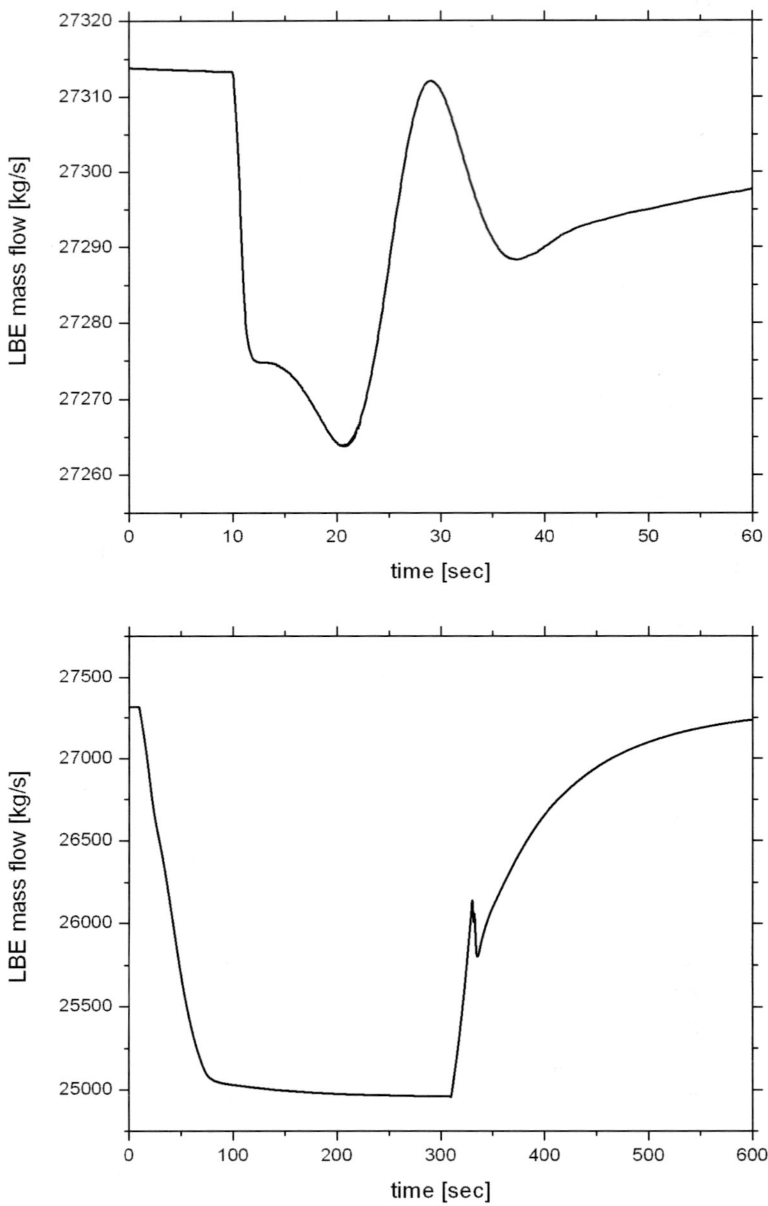

Figure 5.20 LBE mass flow behavior caused during a beam interruption of 1 second (top) and 5 minutes (bottom) in a $450MW_{th}$ two-phase flow reactor

Loss of Heat Sink

The initial condition for this transient is the reactor steady-state full power operation. At time *t=5 seconds,* the heat exchanger feedwater pump fails and heat removal in the heat exchanger stops. The control unit stops the proton beam and the core power drops immediately. Then, heat removal through reactor vessel auxiliary cooling systems (RVACS) is started. The detailed description of events following a loss of heat sink were previously mention in section 5.2. However, the detailed model provides a more accurate calculation of the reactivity feedback indicating that the subcritical conditions are assured as shown in figure 5.21.

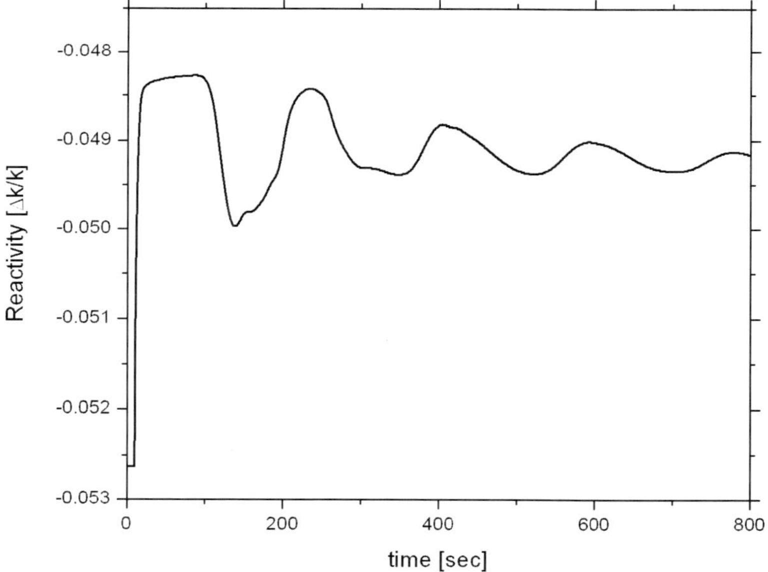

Figure 5.21 Reactivity behavior during a LOHS in a 450MW$_{th}$ two-phase flow reactor

Figure 5.22 shows the behavior of cladding temperature at different axial positions in the central zone. The results show a similar oscillatory pattern to that presented in section 5.2. This is caused by the density changes in the gas lift, which originate from the coolant temperature changes. The buoyancy force is affected by these fluctuations and its effect is the induction of flow oscillations.

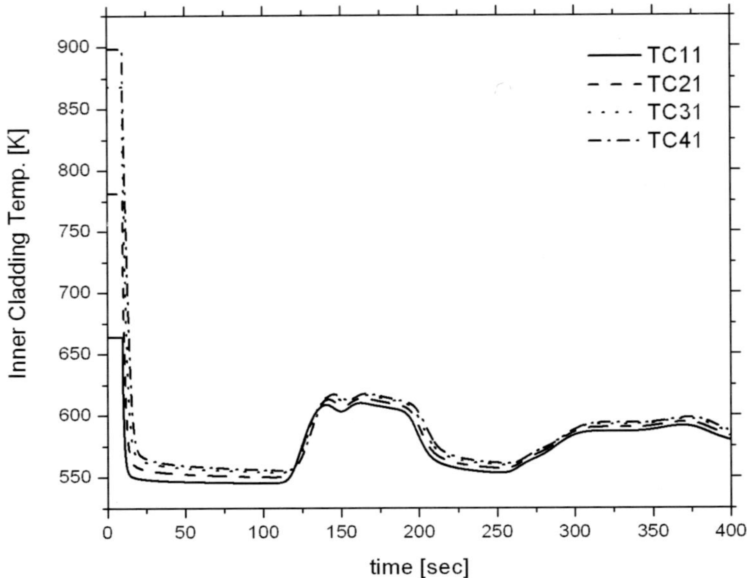

Figure 5.22 Inner cladding temperature during a LOHS in a 450MW$_{th}$ two-phase flow reactor

Unprotected Loss of Heat Sink

The initial conditions for the transient analysis are the same of the loss of heat sink transient, but in this particular, the beam is not shutdown. The consequences of an unprotected loss of heat sink transient could be core melting and/or possible re-criticality accidents by fuel displacement.

When the heat exchanger stops, the coolant from the riser flows inside the downcomer still hot, the mass flow by thermal gradients is reduced and the fuel temperature increases. As a result, the Doppler negative reactivity feedback reduces the core power, but the presence of the neutron source moderates the feedback effect. The total power reduces to 92% during the first 5 minutes of the transient. As a consequence of coolant mass flow rate reduction, the fuel pins start to heat up. The inner cladding temperature of fuel pins located in a central region reaches the limiting temperature value of 1000 K after the first 200 seconds (see figure 5.23). One could presume that there is enough time for the operator and/or control systems to perform shutdown actions with active and passive mechanisms in order to stop the neutron source.

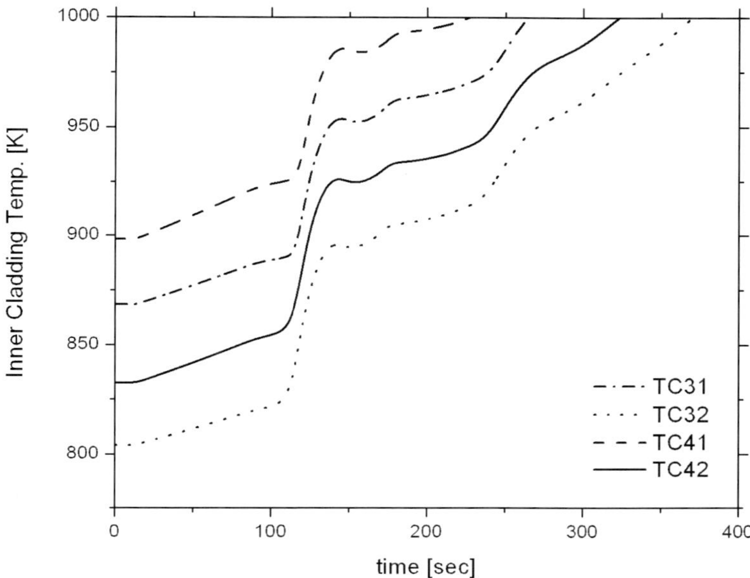

Figure 5.23 Inner cladding temperature during a ULOHS in a 450MW$_{th}$ two-phase flow reactor

Unprotected Loss of Gas Lift

The initial condition of this transient is the steady-state full power operation. Then at *t=10 seconds*, a sudden shutdown of gas lift pump occurs. Meanwhile, the proton beam is not shutdown. The loss of gas lift causes a reduction on coolant mass flow by buoyancy decrease. The effect is a fuel temperature increase and consequent Doppler negative feedback.

When the gas in the riser disappears, the reactor seeks a new stationary condition for operating as a single-phase flow reactor. This takes a couple of minutes for a new steady-state condition to be found. In the meantime, the core heat removal is diminished and despite the heat exchanger and RVACS are still working, the maximum cladding limit temperature is crossed at some locations in the central-top core zones (see figure 5.24).

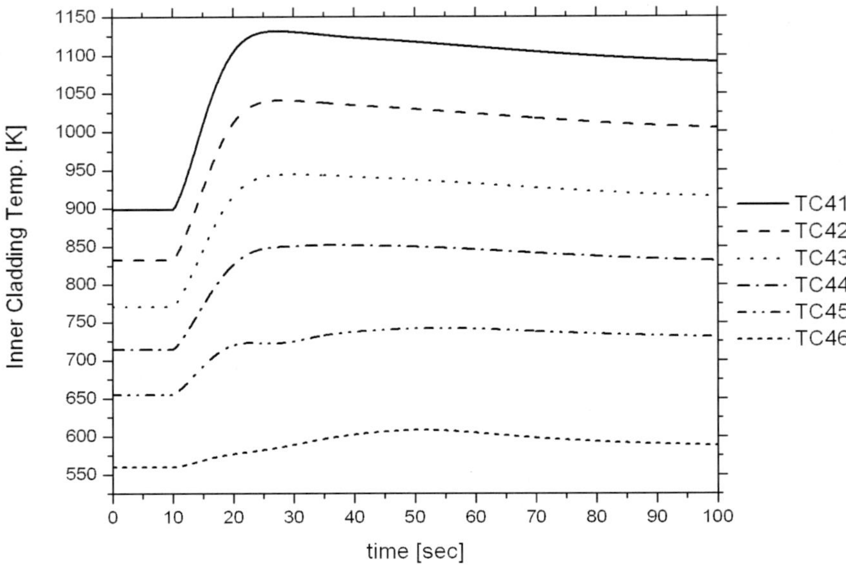

Figure 5.24 Inner cladding temperature during a ULOG in a 450MWth two-phase flow reactor

5.4 Comparison of Thermalhydraulic Models

The comparison between 1D and 2D thermalhydraulic modeling approaches is an interesting added to the present analysis and could be useful to confront their competitive advantages for further use. One should realize that the 2D approach is capable of reproducing the exact power shape in the reactor core and the hydrodynamic details of the reactor pool. On the other hand, the difference in maximum power at steady operation, which was already discussed in chapter 4, demonstrates the necessity of detail modeling even at the start of a conceptual design.

For the ADSR, the power distribution shape is very steep towards the center of the core, hence the coolant outlet temperature significantly varies from fuel assemblies located near the central region and fuel assemblies at the outer zones. Nevertheless, the average coolant core outlet temperature is very similar as it was shown in table 4.3. When transient results are compared, the result indicates that a 1D model is more conservative with regards to positive neutronic feedbacks. Figure 5.25 shows the positive reactivity insertion caused by the fuel temperature decrease after a beam trip. In a 2D model, the fuel average temperature is influenced by power distribution shape, the low temperature values at the outer zones reduce the absolute magnitude of temperature difference. Thus the net effect on reactivity feedback is lower.

If we look at the same issue from a different perspective, the 1D model is less conservative regarding to power excursions. The reasons are opposite to those presented above. When the reactivity feedback is larger, the calculated power levels are less reliable and the predicted cladding and fuel temperatures are under estimated. Another significant difference is the net effect of the buoyancy force given by the two-phase flow. Figure 5.26 shows a marked difference in the percentage of mass flow due to buoyancy by thermal gradients and buoyancy by two-phase fluid mixture. The lower average temperature obtained when spatial effects are included makes the 2D-axisymmetrical model highly driven by buoyancy by two-phase fluid mixture, while the comparison with the 1D model shows that the buoyancy from the thermal gradients is larger.

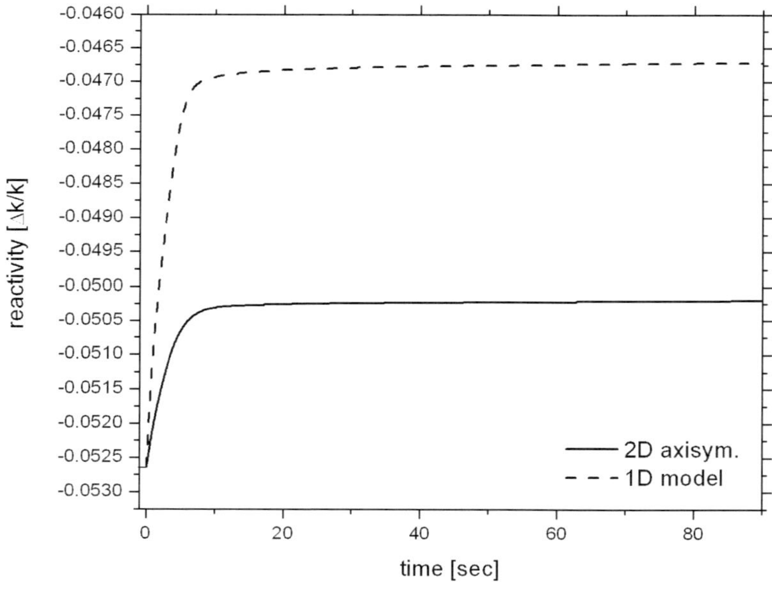

Figure 5.25 Comparison of reactivity feedback during a beam trip transient using the 1D and 2D modeling approaches in a 650MW$_{th}$ two-phase flow reactor

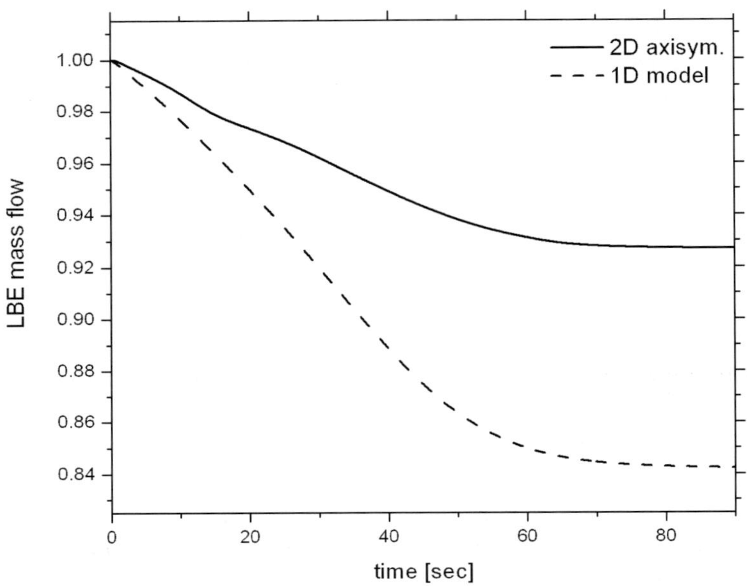

Figure 5.26 Comparison of LBE mass flow behavior during a beam trip transient using the 1D and 2D modeling approaches in a 650MW$_{th}$ two-phase flow reactor

5.5 Concluding Remarks

Transient analysis have shown that the two-phase enhanced natural circulation of 450 MW$_{th}$ is safe for all transients except the unprotected loss of heat sink and unprotected loss of gas lift. The cladding temperature limit has been reached after some minutes in the ULOHS and after some seconds in the ULOG, but this time could be sufficient to allow additional actions for proton beam shutdown. Additionally, some of the calculated transients show that the cladding and structures could be subjected to high cycling thermal loads. The control of cooling rates during transient events is required to avoid high thermal gradients, which threaten the reactor core and internal structures. The high thermal gradients are likely to appear due to the existence of a large temperature difference of the coolant across the core, which is a requirement to achieve high thermal efficiencies in liquid metal reactors.

" To hasten the day when fear of the atom will begin to disappear from the minds, the people and the governments of the East and West, there are certain steps that can be taken now"

Mr. D. Eisenhower – Atoms for peace, 8 dec 1953

6 Structural Integrity Assessment of Fuel Pin Cladding

Failure modes progressing in the fuel pin cladding are caused by mechanical loads and high temperatures, which create elastic and plastic deformation. Furthermore, the heating and cooling cycles during start-up and shutdown operations, cause thermal stress cycling that combined with the inner gas pressure of the fuel pin, produce thermo-mechanical fatigue. To estimate their effect, a thermal-structural analysis has been performed for the fuel pin cladding. The model verifies through inelastic finite elements calculations if the design can withstand the load conditions.

6.1 Introduction

The results obtained from the transient analysis have demonstrated that the ADSR cooled by buoyancy driven two-phase flow exhibits a reliable and safe response during most of the transients. The unfavorable scenarios are the unprotected transients, yet the buoyancy driven concept offers enough grace time for the operator or control systems to shutdown the neutron source. In terms of controllability, the subcritical state is always conserved and a careful core design could assure good margins to criticality. Severe accidents involving massive positive reactivity insertions or core disruption could become a source of criticality accidents, but they were not considered in the present study. In contrast, the determination of integrity losses that leads to severe accidents becomes an interesting aspect to look at. Consequently, the structural integrity assessment recognizes the ADSR ability to resist the cyclic loading conditions.

Up to now, the safety analysis has shown the existence of a large temperature gradient from core inlet to outlet. The gradient is necessary to achieve a high thermal power output and high efficiency, but may become a threat to structural components in conditions of thermal stress cycling. Likewise, during load periods, the reactor core and internals experience elevated temperature conditions, where structural strength is influenced by several time-dependent mechanisms such as oxidation, creep and phase transformation (Mannan and Valsan, 2005). All these effects reveal the need for restrictions to lifetime to avoid loss of integrity. For this reason, life evaluation tests of materials have been developed at low cycle rates and temperature hold times (Sauzay et al., 2004; Lu et al., 2005). The experiments demonstrate that the fatigue life is affected by the hold times since, the number of cycles to failure decreases when the hold time is increased. In the same way, the number of cycles decreases as the test temperature is increased. Therefore, low cycle fatigue is an important consideration in the design of high temperature systems subjected to thermal transients (Mannan and Valsan, 2005).

For the ADSR, the high beam trip frequency of the accelerator (1 trip/hour) imposes a high number of cycles and this could extremely degrade the material strength. To study the reliability of the ADSR concept, a thermal-structural stress transient analysis has been performed. The analysis focuses on the fuel pin cladding since it is one of the most critical components. The failure of the cladding implies the failure of the first containment barrier. The consequence is the release of radioactive material contaminating the coolant and the reactor pool. Additional consequences could be a positive feedback from voids in the core, pin buckling and cladding removal, which could evolve in a core blockage or a core disruption accident.

The method of assessment is based on an inelastic finite element thermal-structural analysis with standard regulations established for these problems. These standards, like the French RCC-MR, the German KTA and the American ASME code section III, apply to nuclear power plants items in general, and sometimes specifically considering LMFR. All methods exclude irradiation and corrosion effects (Bauermeister and Laue, 1989). The present analysis has followed the criteria presented by the ASME code case N47; rules for elevated temperature service. It should be noted that the reason for this choice is merely the availability of literature. All codes present similar results, despite the approach sometimes is rather different.

The work fulfills a detailed inelastic analysis, which avoids the conservative results from the elastic or the simplified inelastic methods developed for simplification. The present chapter is divided in 3 sections: the first section introduces the physical model for thermal-structural transient analysis and creep in a fuel pin. The description of the heat conduction model is omitted as a smilar model has already been described in chapter 3. Following the model descriptions the assessment rules are listed. Then, the stress-strain results from thermo-mechanical transient analysis are presented and the evaluation rules are applied. Consequently, the creep-fatigue damage interaction is quantified to draw conclusions about structural integrity of the fuel pins during their lifetime.

6.2 Thermal-Structural Cladding Model

Fuel pin performance is affected during its lifetime by several factors such as linear power, fuel and cladding creep, embrittlement, swelling and annealing, fuel cracking, fission gas release, irradiation damage and differential thermal expansion (Waltar and Reynolds, 1981). During design all these physical phenomena are combined in a thermal-stress analysis. Dedicated design codes have been created for this purpose, which are capable of integrating these inter-related problems in a detailed time-dependent calculation of the fuel pin lifetime operational cycles.

The present analysis considers that when the fuel assembly is placed in the reactor core, it has already been pressurized. Once the reactor has been in operation, a radial temperature gradient across the cladding is imposed. This temperature difference from inner and outer side of the cladding induces thermal stress. Figure 6.1 shows the loading type applied to the cladding during operation. If a transient occurs, the heat deposited in the fuel changes and hence, the change in radial temperature gradient induces compression and tension stresses. The cyclic loads induce fatigue stresses on the cladding. Additionally, the elevated temperature condition in fuel pin cladding subjects the material to creep. The other previously mentioned phenomena are disregarded, nevertheless one has to be aware that these effects could worsen the loading scenario.

Model Description

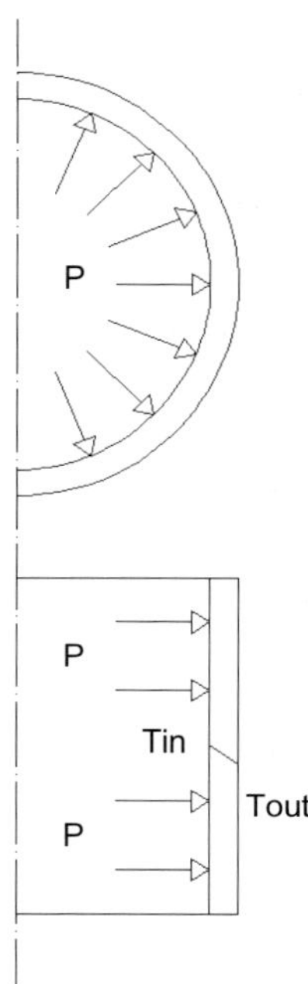

Figure 6.1. Description of loads for the cladding model (top and front views of a cladding section)

The fuel pin is a thin-walled cylinder deformed by symmetrically applied loads (pressure and radial temperature gradient). The model proposed is an axisymmetric solid representation of a cladding section as it is shown in figure

6.1 (bottom). The model is built in the commercial finite elements program COMSOL Multiphysics. The model connects transient heat conduction across the cladding with the transient thermal-structural stress including creep. The loads are the fission gas inner pressure and the temperature gradient across the thickness given the inner and outer cladding temperatures. These values were calculated during the transient analysis described in section 5.3 and obtained for 24 different locations in the core.

The model assumptions are:

- Since the cylinder length is very large compared to the thickness, one can study a piece of cladding far way from the ends of the cylinder and hold in equilibrium with its neighboring sides by means of bending moments (*Principle of Saint-Venant)*
- The model is long enough that the bending effects of the edges, do not alter the axial mid-plane and hence, it represents local stress-strain conditions of the cladding under the specified loading
- The temperature is assumed to be independent of the axial coordinate

Thermo-mechanical Stress Model

The strain at a point can be decomposed into elastic strain ε_{el}, thermal strain ε_{th}, and creep strain ε_c.

$$\boldsymbol{\varepsilon} = \boldsymbol{\varepsilon}_{el} + \boldsymbol{\varepsilon}_{th} + \boldsymbol{\varepsilon}_c \tag{6.1}$$

and the symmetric strain tensor consists of the normal and shear strain components

$$\varepsilon = \begin{bmatrix} \varepsilon_r & \gamma_{r\theta} & \gamma_{rz} \\ \gamma_{\theta r} & \varepsilon_\theta & \gamma_{\theta z} \\ \gamma_{zr} & \gamma_{z\theta} & \varepsilon_z \end{bmatrix} \tag{6.2}$$

In the same way, the stress in the material is described as

$$\sigma = \begin{bmatrix} \sigma_r & \tau_{r\theta} & \tau_{rz} \\ \tau_{\theta r} & \sigma_\theta & \tau_{\theta z} \\ \tau_{zr} & \tau_{z\theta} & \sigma_z \end{bmatrix} \tag{6.3}$$

The stress-strain relation for linear conditions written in vector form writes as

$$\boldsymbol{\sigma} = \boldsymbol{D}\left(\boldsymbol{\varepsilon} - \left[\boldsymbol{\varepsilon}_{th} + \boldsymbol{\varepsilon}_c\right]\right) \tag{6.4}$$

where D is the elasticity matrix function of the Young's modulus E and the Poisson's constant ν. The thermal strain is defined with respect to a reference temperature free of stress. The solution implementation is based on the equilibrium equations expressed in the global stress components. The stress components are independent of θ and the shear stresses $\tau_{r\theta}$ and $\tau_{\theta z}$ vanish on account of the symmetry. The equilibrium equations are

$$\frac{\partial \sigma_r}{\partial r} + \frac{\partial \tau_{rz}}{\partial z} + \frac{\sigma_r - \sigma_\theta}{r} = F_r \tag{6.5}$$

$$\frac{\partial \tau_{rz}}{\partial r} + \frac{\partial \sigma_z}{\partial z} + \frac{\tau_{rz}}{r} = F_z \tag{6.6}$$

where F are body forces. The elasticity matrix takes the following form

$$\boldsymbol{D} = \frac{E}{(1+v)(1-2v)} \begin{bmatrix} 1-v & v & v & 0 \\ v & 1-v & v & 0 \\ v & v & 1-v & 0 \\ 0 & 0 & 0 & (1-2v)/2 \end{bmatrix}$$

The displacement components in radial and axial directions are denoted as $\boldsymbol{u}$ and $\boldsymbol{w}$. The expressions for strain components are written as

$$\begin{bmatrix} \varepsilon_r \\ \varepsilon_\theta \\ \varepsilon_z \\ \gamma_{rz} \end{bmatrix} = \begin{bmatrix} \dfrac{\partial u}{\partial r} \\ \dfrac{u}{r} \\ \dfrac{\partial w}{\partial z} \\ \dfrac{\partial u}{\partial z} + \dfrac{\partial w}{\partial r} \end{bmatrix}$$

The substitution of these strains in the equilibrium equations, leads to two partial differential equations of second order for the variables, ***u*** and *w*. The problem is then reduced to the solution of these two equations by a numerical finite element method.

Creep Model

The creep strain rate is a function of three variables: the stress, the temperature and time. The relations can usually be separated as

$$\dot{\varepsilon}_c = \dot{\varepsilon}_c(\sigma, T, t) = f_1(\sigma) f_2(T) f_3(t) \tag{6.7}$$

Experimental work has demonstrated that some of the effects can be fitted to general functions, such as the Arhenius law for temperature effects, power laws for time and stress effects. For the present analysis, "The Norton Law" is selected and applied assuming creep during isothermal conditions as

$$\dot{\varepsilon}_c = A f_1(\sigma) = A\sigma^n \tag{6.8}$$

where A and n are constants function of the material properties.

In the multiaxial generalization, the effective Von Mises stress $\bar{\sigma}$ is used to calculate the creep strain rate. This relation is proportional to the stress deviator s_i and its expression takes the form

$$\dot{\varepsilon}_{c_i} = A \frac{f(\bar{\sigma})}{\bar{\sigma}} s_i \tag{6.9}$$

where the stress deviator s_i is defined as

$$s_i = \sigma_i - \sigma_v \tag{6.10}$$

and σ_v is the hydrostatic pressure defined as

$$\sigma_v = \frac{1}{3}(\sigma_1 + \sigma_2 + \sigma_3) \tag{6.11}$$

The value of n coefficient equals 4.5 and was taken for a SS304 from Skrzypek (1986). The constant A is obtained from figure T-1800A4 (SS304) in ASME Code

Case N47 (see figure 6.2). This table exhibits that SS304 experiences a strain of 1% when is loaded with stress of 100 MPa at 790 K; the values were estimated for 5 years of loading time (43800 hours). The experiment was reproduced with a finite element model and the constant A scaled to fit the results listed by ASME for the creep test time T. The final expression has the following form

$$\dot{\varepsilon}_{c_i} = \frac{3}{2\mathrm{T}}\left(\frac{\bar{\sigma}}{\sigma_c}\right)^{n-1}\frac{s_i}{\sigma_c} = A(\bar{\sigma})^{n-1}(s_i) \tag{6.12}$$

with A equal to

$$A = \frac{3}{2T\sigma_c^n} \tag{6.13}$$

This method is analogous to the one applied by Becker to study thermally induced creep in a hollow sphere and is presented in understanding non-linear finite elements analysis through illustrative benchmarks (2001).

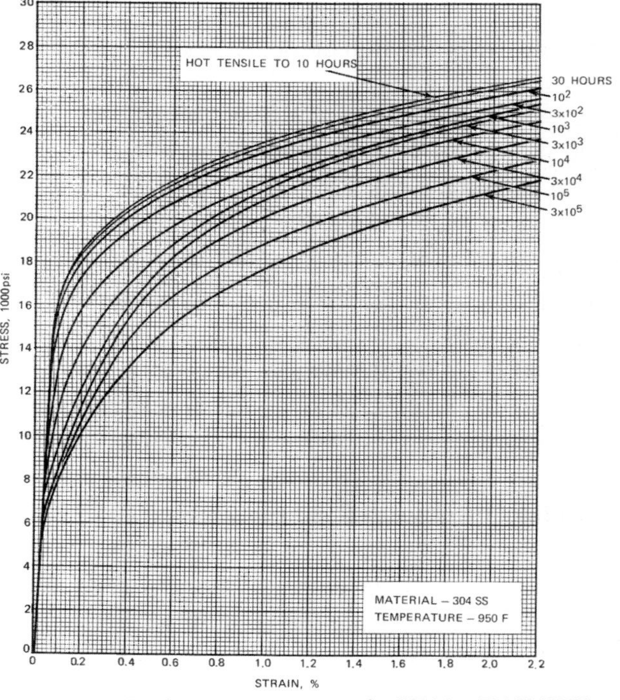

Figure 6.2. Isochronous creep curve for SS304 at 790 K (950F) (source: ASME pressure vessel code Case N47)

Chapter 6

Assessment Rules

According to ASME code case N47, the combination of loadings service shall be evaluated for accumulated creep and fatigue damage including hold time and strain rate effects. For a design to be acceptable, the creep and fatigue damage should satisfy the following condition:

$$\sum_{j=1}^{p}\left(\frac{n}{N_d}\right)_j+\sum_{k=1}^{q}\left(\frac{\Delta t}{T_d}\right)_k\leq D \tag{6.14}$$

where n is the number of applied cycles of loading conditions of type j and N_d is the number of cycles from one of the design fatigue curves corresponding to the maximum metal temperature during the cycle j. Thus $[n/N_d]$ is the fatigue damage fraction. In the same way, k is the number of intervals under a stress-temperature combination. Then, Δt is the time duration of the load conditions during an interval k and T_d the allowable time duration determined from a stress to rupture curve at the given loads stress-maximum temperature. Thus $[\Delta t/T_d]$ is the creep damage fraction. The total damage must not exceed the creep-fatigue damage envelope D for the material.

Due to the nature of the loading, inelastic strains have to be determined. The ASME code proposes the determination by either a detailed inelastic analysis or simplified inelastic method. Simplified inelastic analysis is a conservative approach that facilitates the multiple analyses in design processes, providing quick answers to enable designers to modify the design or material selection. Although simplified inelastic analysis can be performed, the present analysis has preferred the use of a classical treatment based on the Von Misses effective stress to consider a creep strain rate, as it was illustrated in the previous section. Still, it has the drawback of the higher computational effort but the advantage of an increased accuracy.

Creep Damage Fraction

For each loading interval k there is a maximum effective stress S_k. This value is divided by a safety factor K' material dependent and used to find the reference time T_d in the expected minimum stress to rupture curves.

There are also rules of design against gross distortion and fatigue applied strain and deformation limits. These rules define that the maximum accumulated inelastic strain must not exceed the following values:

- Service levels "normal, upset and faulty" – strains averaged through the thickness 1%, strains at the surface due to an equivalent linear distribution through the thickness 2%, and local strains at any point 5%
- Service level "emergency" – not applicable except as necessary to satisfy functional requirements

These limits apply to the maximum positive value of the three principal strains accumulated over the expected operating lifetime and computed for a steady-state period during which significant transients have not occurred.

Fatigue Damage

The equivalent strain range ε_t is used to evaluate the fatigue damage for the loading cycles. The equivalent strain range is computed by determining the principal strain versus time for the cycle. At each time interval the main strain differences are determined

$$\varepsilon_{12} = \varepsilon_1 - \varepsilon_2 \qquad \varepsilon_{23} = \varepsilon_2 - \varepsilon_3 \qquad \varepsilon_{31} = \varepsilon_3 - \varepsilon_1$$

and the history of the change in strain differences by subtracting their values in time

$$\Delta(\varepsilon_1 - \varepsilon_2) = \varepsilon_{12} - \varepsilon_{12i} \quad \Delta(\varepsilon_2 - \varepsilon_3) = \varepsilon_{23} - \varepsilon_{23i} \quad \Delta(\varepsilon_3 - \varepsilon_1) = \varepsilon_{31} - \varepsilon_{31i}$$

Then, the equivalent strain range is computed as

$$\varepsilon_t = \Delta\varepsilon_{equiv} = \frac{\sqrt{2}}{3}\left\{\left[\Delta(\varepsilon_1 - \varepsilon_2)\right]^2 + \left[\Delta(\varepsilon_2 - \varepsilon_3)\right]^2 + \left[\Delta(\varepsilon_3 - \varepsilon_1)\right]^2\right\}^{1/2} \tag{6.15}$$

and this value is used to enter the fatigue curves from the ASME code to determine N_d for the corresponding maximum metal temperature.

6.3 Assessment of Fatigue-Creep Damage in the Fuel Pin Cladding

In view that a large amount of beam interruptions are expected to occur if the current accelerator technology is driving the ADSR, the creep-fatigue damage analysis is focused on the effects of beam trips on cladding lifetime. Since the LANSCE accelerator produces high-energy particles, high power and has been well documented regarding the reliability of the beam (Eriksson, 1998), we have decided to use it as reference to the present analysis. Eriksson has reported a detailed description of accelerators' fault frequencies and time duration in the LANSCE facility of LANL in the U.S. Table 6.1 shows the quantity of beam interruptions in a measured time interval determined by Eriksson (1998) for the LANSCE accelerator. If a direct proportionality relation in time is assumed, the data can be extrapolated to the ADSR operation time of 1 year.

Table 6.1. Number of beam interruptions determined in the LANSCE accelerator and extrapolated values to the ADSR analysis

Time		LANSCE	ADSR
Total operation time	[hr]	2821	8760
Total dead time	[hr]	381	1183
Total effective time	[hr]	2440	7577
0.5-2 min	[min]	1333	4139
2-5 min	[min]	555	1723
5-15 min	[min]	217	674
15-60 min	[min]	124	385
1-5 hr	[hr]	53	165
> 5 hr	[hr]	13	40

Table 6.1 shows the need of overcoming a total of 7126 beam interruptions during one year. The interruption could last from 30 seconds to 5 hours until the proton beam current is restored. Such rate would cause cladding damage earlier than expected in a conventional reactor, which is designed to withstand a very limited number of scrams (1 or 2 per year). For the ADSR, one should expect a similar result.

Modeling Results

The correctness of the FEM model was checked against steady analytical solutions for a cylinder with internal pressure and similarly, for a cylinder with a radial temperature gradient. The modeling result has shown a good agreement with the correlations proposed by Timoshenko (1951) for this type of loading. Likewise, the FEM creep model was benchmarked performing an uniaxial tension creep test and its results compared against the strain values reported in the isochronous curve from ASME code, calculated for the same material, same temperature condition and hold time.

The cladding is loaded with a positive inner pressure of 1 MPa and a radial temperature gradient. The applied pressure is the net pressure applied over the cladding wall if inner pressure is 2 MPa and the hydrostatic pressure 1 MPa. The temperature gradient corresponds to the radial temperature profile through the cladding thickness at the specified core position. This is the core zone 31 (following the nomenclature described in chapter 5). It was selected after a careful comparison based on the results from the transient analysis described in section 5.3. The inner and outer cladding temperatures at steady state conditions are 868.32 K and 830.16 K respectively, and the gradient fits a logarithmic profile.

Creep Results

The results from a steady state thermal-structural analysis are indicated in figure 6.3. The loading conditions are inner gas pressure and the radial temperature gradient, which is hotter at the inner side and colder at the outer side. The results indicate that the hoop and axial stresses are negative (compression) at the cladding inner side and positive (tension) at the outer side.

Figure 6.4 shows how the stress relaxes in time by creep strain effects. We have preferred to plot Von Mises stress, considering that creep theory uses the effective stress to determine the creep strain rate (equation 6.9). The calculation is performed for 3 years lifetime, in which the suggested fuel burn-up is achieved (MacDonald and Buongiorno, 2001). The results are mainly illustrative of the creep phenomenon under the loading conditions, since the loading cycles for the cladding are of short duration. It is expected that there will be annual stops for inspection, refueling and maintenance tasks in which the material recovers partially.

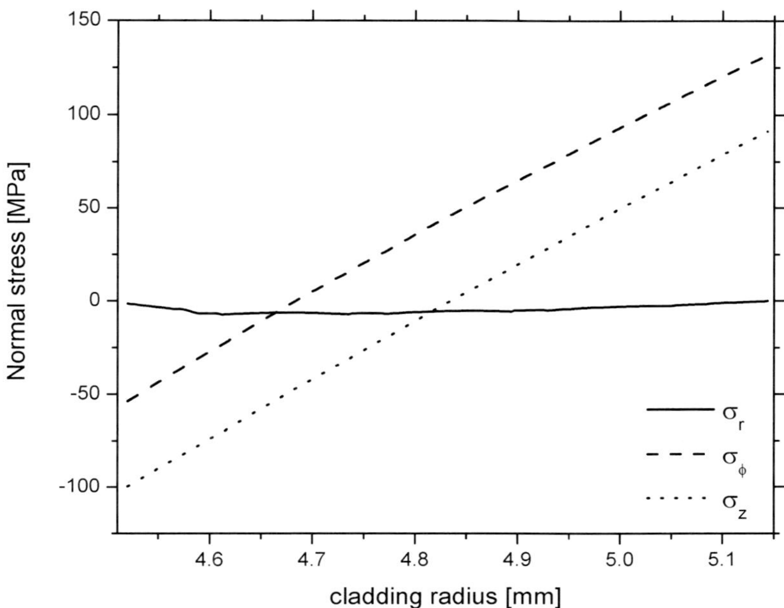

Figure 6.3. Stress induced by inner pressure and temperature gradients across the cladding

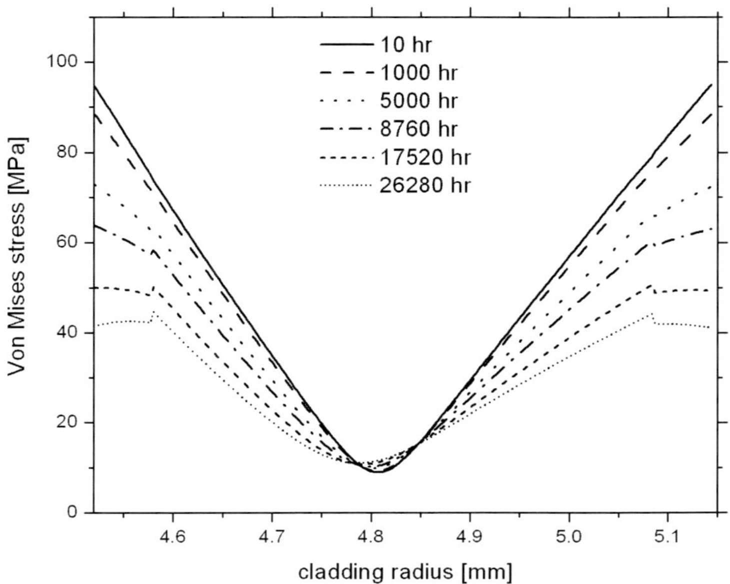

Figure 6.4. Relaxation of Von Mises stress in cladding cross section and due to creep behavior for 3 years time operation

The creep strains have shown that the strain range does not exceed 0.1% during a time interval in steady operation. The creep damage fraction is obtained with the minimum time to rupture T_d in figure I-14.6A (ASME code case N47) given a Von Mises applied stress. The Von Mises applied stress at the surface is around 100 MPa thereby the correspondent minimum rupture time is 10.000 hours. Consequently, if the amount of beam interruptions per year is 7126, the effective operating time is 7577 hours and the total time per year is 8760 hours, the damage fraction equals 0.61.

Fatigue Results

Different transient scenarios were modeled such as the beam trip, a loss of heat sink and different beam interruptions as shown in chapter 5. These transients are likely to occur during the ADSR lifetime and that is why they were chosen. The starting conditions for modeling thermal-structural transients are the state of stress given by the loads at steady-state operating conditions. Then, the thermal-stress evolves as the temperature gradient in the cladding changes.

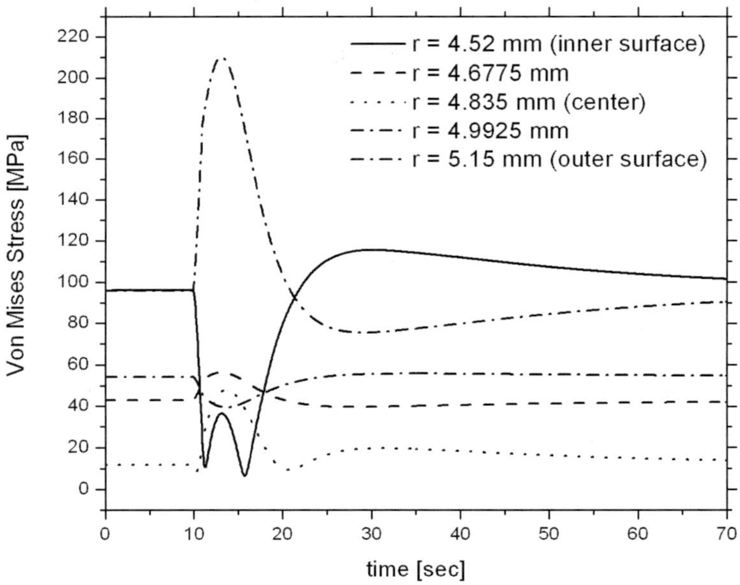

Figure 6.5. Von Mises stress in time for 5 points in the cladding thickness, during a beam interruption of 1 second

Figure 6.5 shows the Von Mises stress during a beam interruption of 1 second for 5 equidistant points across the cladding thickness. The result shows that the highest peak stress occur at the outer cladding surface. At this side of the cladding, there is thermal shock produced when the power drops. On the opposite side, the Von Mises stress reduces at a similar rate because tension is released. After 1 second, the beam power returns and heat is deposited in the fuel. The inner cladding temperature increases, the initial temperature gradient is recovered and sustained. Meanwhile, the stresses are relaxed back to the initial state.

Figure 6.6 shows the result of a similar transient in which the beam interruption lasts 1 minute. In this case, the Von Mises stress at the inner cladding surface develops further until a maximum peak. Then, it is relaxed slowly by the new evenly distributed temperature gradient. When the beam power is back, the cladding experiences a newly stressing condition. The maximum peak is found at the inner side of the cladding since the thermal shock happens there now. It is important to note that any beam interruption longer than 15 seconds have already involved the maximum peak stress. Also, the relaxation time implies more than 100 seconds to get back into the original state of stress.

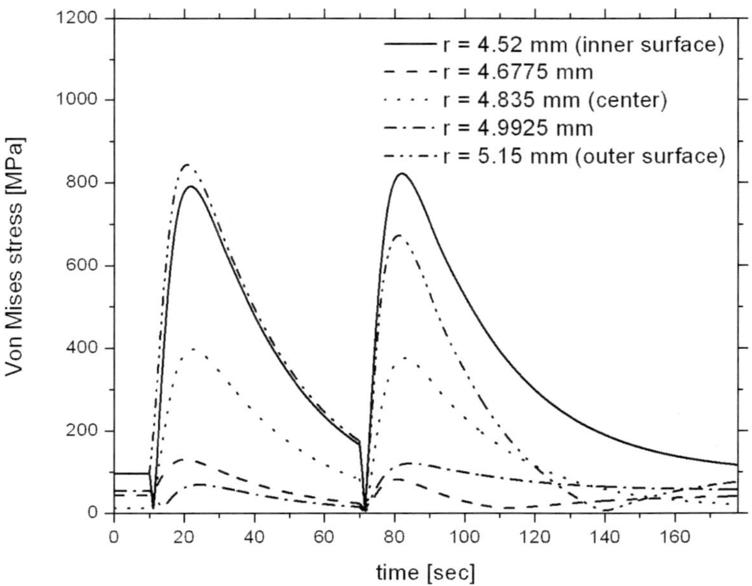

Figure 6.6. Von Mises stress in time for 5 points in the cladding thickness, during a beam interruption of 1 minute

To illustrate the cladding elastic response during the beam interruption of 1 minute, the principal strains at the outer surface are plotted in time. Figure 6.7 shows two peak strains (positive and negative) in the 0.6% range. They occur at the beginning of the transient since the thermal shock induces high strains rates by material contraction. The deformation is reflected in the increasing effective Von Mises stress as seen in figure 6.6. Once the temperature gradient is uniform, the deformation reduces and the stress relaxes. When the beam returns, new deformation is originated, but the strain direction swaps, as the cladding is experiencing expansion.

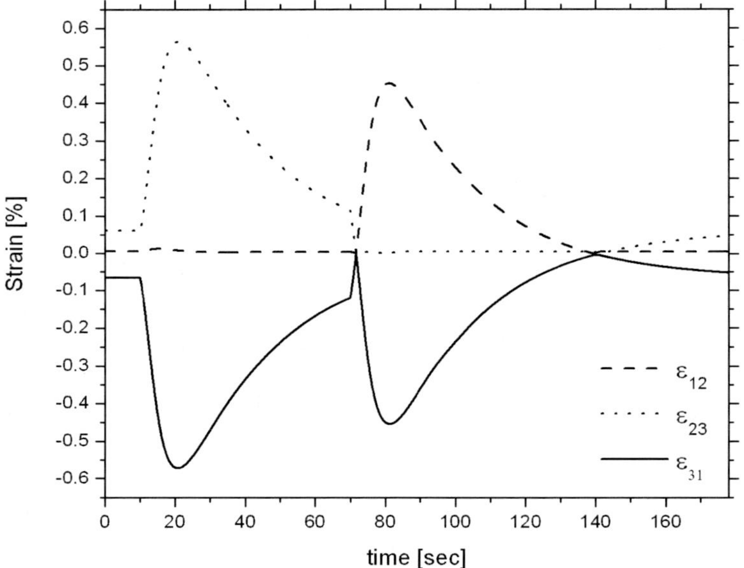

Figure 6.7. Main strains in time at a single point on the outer cladding surface, during a beam interruption of 1 minute

Additional calculations have shown that the stress-strain produced during a loss of heat sink transient or a sudden beam shutdown lead to similar result of the 1 minute beam interruption. The peak stresses are caused by thermal shock following the beam trip and return. The effects of coolant temperature changes are low since the flow speed is very low.

The equivalent strain range was calculated for the transient beam interruption of 1 minute. The maximum equivalent range is found at the inner cladding surface and has a value of 3.5×10^{-3}. The maximum number of cycles allowed N_d according to table T-1420-1A from ASME is 3.000 cycles. Given these values, the fatigue damage fraction is 2.3. The total creep-fatigue damage fraction is the sum of the individuals creep and fatigue damage fractions. This value is by far out of the damage envelope from ASME, which indicates that the cladding fails rapidly during the first year of operation.

The effect of cladding thickness in the stress level was investigated by changing the outer radius of the cladding and computing the same transients. The result indicates that a thicker cladding (outer radius: 6mm) holds higher thermal inertia and therefore, the peak stresses are larger. For this reason, the numbers of fatigue cycles and time to rupture are reduced and hence, the damage fraction is increased. In a short beam interruption the stress level does not change significantly, whereas in the beam interruption of 1 minute the peak stress increases by 25% when the cladding thickness is doubled. The relaxation time is longer since the thermal inertia is larger as well.

6.4 Concluding Remarks

The structural integrity assessment method proposed by the ASME code case N47 has been applied to the fuel pin cladding. The result indicates that the high temperature operating conditions and the thermal cycling caused by beam interruptions will considerably reduce the lifetime of fuel pins. Having evaluated the fatigue and creep damage separately, the total damage fraction lies far away of the interaction envelope from ASME code. Even, if a non-conservative boundary is used, the fatigue damage fraction is exaggeratedly large. The effect of beam interruptions can be distinguished in two main time scales, the short beam interruptions, which do not reach the largest peak stress, and the long beam interruptions, which exhibit double maximum peaks. For the case studied, it is in the range of the 12 to 15 seconds of interruption time.

"Imagine a world where every human being would live in freedom and dignity, where we would settle our differences through diplomacy and dialogue and not through bombs or bullets, where the only nuclear weapons remaining were the relics in our museums, imagine the legacy that we could leave to our children"
Mohamed ElBaradei, during his speech at the Nobel prize ceremony 2005

7 Conclusions and Recommendations

This thesis has presented a multidisciplinary design approach to study the safety aspects of an ADSR. Different physical systems such as neutronics, thermalhydraulics and mechanical components were integrated in a computational model. The model has evaluated the reactor's ability to remain within the safety envelope. Safety analysis was used to assess the performance of the reactor in terms of controllability and integrity. The model has also estimated the passive safety capabilities of natural circulation and enhanced gas lift systems for lead-bismuth cooled reactors.

Regarding the ADSR safety competences, the results indicate that the subcriticality level is remarkably effective to overcome transients involving criticality risk such as positive feedbacks and positive reactivity insertions. The reactor remains subcritical indifferent whether the beam has been shutdown or has not. In contrast, the same transients have a large effect on fuel pin temperatures if the beam is not shutdown. A large percentage of the power remains on, while the coolant mass flow rate stays constant or reduces. Furthermore, the subcriticalilty makes the reactor highly insensitive to negative feedbacks, complicating the shutdown task.

The nature of natural circulation was also studied. The mass flow rate is a function of balancing buoyancy with pressure drops. The coolant capability to develop natural circulation does not depend on one physical property but on the combination of physical properties and geometrical parameters in the non-dimensional Stanton and friction numbers. However, the limited choice of coolants for liquid metal reactors makes the coolant selection an unpractical approach. Instead, the maximization of core power in a pool type reactor is achieved with a good mechanical design. The design constraints are the vessel height, the core pitch-to-diameter ratio, the heat exchanger configuration, the maximum material temperatures, the spatial power distribution shape in the core and the flow hydrodynamics of the reactor pool. Enhanced buoyancy with a gas lift pump holds part of the ability of the reactor to develop natural circulation but, the passive safety is decreased if compared to single phase natural circulation. Nevertheless, this concept holds an improved power capacity, which could range in the order of few hundred MW_{th}. The constraints are the same and additional boundaries rise from the gas lift pump operation, which is restricted by injection capacity and the two-phase flow pattern stability.

Regarding the structural integrity, the large temperature difference across core inlet and outlet becomes a threat to structural components under transient conditions. This large temperature difference is a necessity to obtain large core power and high efficiency since the coolant does not undergo phase changes. The consequence is the increased probability of structural components exposure to large and fast thermal shocks, creating stress peaks that reduce the material strength. Additionally, high temperature operation implies reduction of materials strength due to creep loads.

Regarding the accelerator's beam trip frequency, the study shows that the feasibility of coupling a particle accelerator with a subcritical nuclear reactor using today's accelerator technology is far from reality. Current accelerators experience around 7000 transient beam interruptions in 1 year. The reliability of the accelerator should be increased by a factor of 5 in order to avoid cladding failure during the first year of operation. This value is indicative for R&D needs. Furthermore, factors such as short beam interruptions (of the order of 0-10 seconds), fuel cycle requirements, other components lifetime (beam window or the core support structure) and damage by corrosion and irradiation could increase this factor.

To implement a reactor safety strategy for the ADSR following the recommendations of a defense in depth concept, one has to consider that the first level of defense should aim to prevent deviations from normal operation and avoid actions that reduce integrity losses. This has to be achieved with passive mechanisms (negative reactivity feedbacks) and a conservative design (selection of a suitable subcriticality level). The accelerator design should aim to reduce the beam trips and the active controls should prevent the beam tripping.

The second level of defense should aim to avoid deviations that abandon certain margin to the safety envelope. This level of defense should be completely supported by active means, it should detect and take actions to control that accident escalation does not evolve. The only possible action to execute effective control when criticality is approached is to shutdown the proton beam. To avoid controllability problems the shutdown action should be fast. In contrast, to avoid integrity losses when there is not a criticality risk, the shutdown should be slow. It is evident that the accelerator's control has to be selective to handle different types of action depending on the transient. The proton beam current needs to be adjusted in order to control the neutron source strength. Control devices should look after measurement of core reactivity and the detection of positive reactivity sources such as, detecting void in the downcomer, void coming from fuel pin failures as well as, detecting low coolant temperature at the core inlet from inhomogeneous or stratified flow patterns. Flow blockages could be detected by measuring coolant temperature changes or pressure drops per fuel assembly.

The third level of defense intercepts the deviations, which were not moderated at the first and second level of defense. This stage is bounded by the safety limits. The actions should be executed by passive means and also consider fast and slow responses as well. This level of defense has to focus on the application of passive systems using moderators and/or neutron absorbers. However, the best action is to assure passive means to stop the proton beam. For these, some proposals are the loss of vacuum in the beam tube by a rupture and/or melting disk, but also thermal-mechanical displacement of the spallation target outside the core region by means of an expanding material.

The fourth level of defense should address the mitigation effects in case that the accident has progressed.

The inherent safety attached to the subcriticality in an ADSR presents large benefits to avoid criticality accidents. However, the development challenges for this technology remain in the core power control and the accelerator reliability. Online reactivity measurement, a variable and reliable beam intensity and the detection and interception of anomalies are the main topics where research activities should be addressed. In other aspects, the technical needs are similar to liquid metal fast reactor technology and their achievements should be easily adopted.

APPENDIX 1

Neutron Kinetic Parameters used by the Point Kinetics Model and Decay Heat Model

Table A1.1. Point Kinetics Parameters
(source: D'Angelo and Gabrielli, 2003; Fomichenko, 2004)

Group, i	Fraction, β_i	Decay Constant, λ_i [s^{-1}]
1	8.60E-05	0.0129
2	7.30E-04	0.0313
3	6.55E-04	0.1346
4	1.27E-03	0.3443
5	5.80E-04	1.3764
6	1.82E-04	3.7425
K effective		0.95
β effective		3.50E-03
λ effective		0.08973
Avg generation time [sec]		2.9E-08
Fuel Temp. coefficient [$\Delta\rho$/K]		-2.319E-05

Table A1.2. Decay Heat Parameters (fitted from source: ANS)

γ_i [s^{-2}]	λ_{i_γ} [s^{-1}]
3.82E-03	2.21E+01
3.01E-03	5.16E-01
1.43E-03	1.96E-01
8.13E-04	1.03E-01
3.26E-04	3.37E-02
1.31E-04	1.17E-02
1.94E-05	3.59E-03
5.46E-06	1.39E-03
4.75E-06	6.26E-04
1.15E-06	1.89E-04
1.91E-07	5.50E-05
4.44E-08	2.10E-05
1.48E-08	1.00E-05
2.93E-09	2.54E-06
1.09E-09	6.64E-07
1.56E-10	1.23E-07
1.32E-11	2.72E-08
4.80E-14	4.37E-09
5.16E-13	7.58E-10
1.48E-16	2.48E-10
1.89E-18	2.24E-13
2.65E-19	2.46E-14
4.39E-19	1.57E-14

APPENDIX 2

Thermalhydraulic Data - Reference Design

Table A2.1. Thermalhydraulics Characteristics
of the Heat Exchanger and Core
(source: Davis et al. 2002; Mac Donald and Buongiorno, 2001)

Core	
power	650 MW_{th}
pitch to diameter ratio	1.2
number of fuel assemblies	157
number of pins per assembly	240
active core length	1.3 m
gas plenum length	1 m
riser diameter	4 m
riser length	10 m
vessel inner diameter	6 m
fuel pin diameter	10.3 mm
pin lattice	Squared

Heat Exchanger	
pipes length	4 m
tube lattice	Triangular
pitch to diameter ratio	1.4
water inlet temperature	200 C
pipe diameter	10 mm

Table A2.2. Material Properties of the Fuel Pin (at 500 C) (source: Mac Donald and Buongiorno, 2001)

Fuel Properties	
Density	12700 kg/m^3
Thermal conductivity	14 W/mK
Heat capacity	350 J/kgK
Radius	4.52 mm

Cladding Properties	
Density	8000 kg/m^3
Thermal conductivity	21.5 W/mK
Heat Capacity	500 J/kgK
Cladding thickness	0.63 mm

Table A2.3. Matrix of Weighting Factors for Power Distribution (adapted from Rubbia *et al.* 1995)

	Radial zone 1	Radial zone 2	Radial zone 3	Radial zone 4	Radial zone 5	Radial zone 6
Axial zone 4	2.23	1.81	1.42	1.08	0.73	0.11
Axial zone 3	3.33	2.61	2.02	1.51	0.96	0.14
Axial zone 2	3.30	2.62	2.03	1.52	0.97	0.14
Axial zone 1	2.18	1.78	1.41	1.06	0.71	0.11

NOMENCLATURE

A_n	cross sectional area
A_{ref}	reference area
C_0	distribution parameter
C_p	heat capacity of the fluid
$C_{p(i)}$	heat capacity of the material in the fuel pin
$C(t)$	delayed power or precursor density in power units
d	diameter of the fuel pins
D_n	hydraulic diameter of the region
D_H	hydraulic diameter of the channel
D	cumulative damage envelope for creep-fatigue
f_n	friction factor
f_{core}	friction coefficient of the core
f_{pipe}	friction coefficient of circular pipes
F	body forces
g	gravitational constant
$G(t)$	correction factor that accounts for effects from neutron absorption
h	heat transfer coefficient of the heat exchanger
h_v	heat transfer coefficient of the vessel
j_g^+	superficial velocity of the gas
k	number of intervals under a stress-temperature combination
k_i	thermal conductivity
K_T	total pressure loss coefficient
K_n	pressure drop coefficients for sudden contractions and expansions
K_μ	friction factor
l	neutron lifetime
l_n	length of different regions
L	reference length
$m_{fuel\text{-}rod}$	mass of the fuel in one rod per zone
n	subscript that indicates the different regions of the loop
n_0	neutron density
n	number of applied cycles of loading conditions of a given type

N_d	number of cycles from one of the design fatigue curves corresponding to the maximum metal temperature during the a load cycle
$N_{rod\text{-}zone}$	number of fuel rods in one zone
$N_{\mu f}$	viscosity number
Nu	Nusselt number
p	pitch between pins
P_{zone}	power in a specified zone
PD_{avg}	average power density of the core
Pe	Peclet number
$P(t)$	prompt power
$P_\gamma(t)$	decay power from fission products
q_n	linear heat flux defined for every region
q_0	external neutron source per unit time
Q	heat source resulting from nuclear fissions
Q_{hx}	heat flux in the heat exchanger
r	radial coordinate
s_i	stress deviator
S_v	perimeter of the vessel
S_{hx}	perimeter of the pipes from the heat exchanger
$S(t)$	external source
St_m	Stanton number
t	time
T	time during uniaxial creep test
T_d	allowable time duration determined from a stress to rupture curve at the given loads stress-maximum temperature
$T(z)$	temperature of the fluid along the loop
T_{ref}	reference temperature of the fluid
T_a	ambient temperature of the vessel surrounding
T_w	wall temperature
$T_f(t)$	average fuel temperature
T_o	temperature reference for zero feedback
u	radial displacement
U	overall heat transfer coefficient
u_{ref}	reference velocity
V_{gj}^+	drift flux
v	velocity
w	mass flow rate

w	axial displacement
w_{ref}	reference mass flow
WF	weight factor of the power distribution matrix
z	axial direction
α_D	Doppler reactivity feedback coefficient
$\alpha(z)$	void fraction
β_T	thermal expansion coefficient of the fluid
β_i	delayed neutron fraction
ε	total strain
ε_t	equivalent strain range
ε_{el}	elastic strain
ε_{th}	thermal strain
ε_c	creep strain
γ	shear strain
λ_γ	fitted parameter from data of fission decay power
λ_i	time constant of delayed group
$\Delta\rho$	difference of density between liquid and gas
ϕ_{fo}	two-phase friction multiplier
Λ	neutron generation time
ρ_c	density of the material
$\rho_{mix}(z)$	density of the gas-liquid mixture in the riser
ρ_{ref}	fluid density referenced at the reference temperature
$\rho_g(z)$	gas density at different positions of the riser
$\rho(t)$	total reactivity
ρ_0	reactivity reference for zero feedback
σ	surface tension of the fluid
$\boldsymbol{\sigma}$	total stress
σ_c	creep stress
$\bar{\sigma}$	Von Mises stress
σ_v	hydrostatic pressure
τ	shear stress
μ	fluid viscosity

BIBLIOGRAPHY

Abderrahim H.A., Kupschus P. & MYRRHA-team, MYRRHA - A Multipurpose Accelerator Driven System (ADS) for Research & Development, Pre-Design Report, R-3595, 2002

Aiello A., Agostini M., Benamati G., et al., Mechanical Properties of Martensitic Steels after Exposure to Flowing Liquid Metals, Journal of Nuclear Materials, Vol 335, pp. 217-221, 2004

Alchagirov B.B., Mozgovoy A.G., Khokonov Kh.B., The surface tension of the near eutectic alloys of lead-bismuth system, Berberov Kabardino-Balkarian State University, 2000

Ambrosini W., Forasassi G., Forgione N., et al., Experimental Study on Combined Natural and Gas-Injection Enhanced Circulation, Nuclear Engineering and Design, Vol. 235, pp. 1179-1188, 2005

ANS standard, Decay Heat Power in Light Water Reactors, Illinois, USA, ANSI/ANS-5.1-1979

Ansaldo Nucleare, XADS PB-Bi Cooled experimental accelerator driven system, Summary Report, ADS 1 SIFX 0500, 2001

ASME Boiler and Pressure Vessel Code, Code Case N-47, American Society of Mechanical Engineers, New York, USA, 1988

Baeten P. and Abderrahim H.A., Measurement of Kinetic Parameters in the Fast Subcritical Core MASURCA, Nuclear Engineering and Design, Vol. 230, pp. 223-232, 2004

Bai H, Thomas B.G., Bubble Formation during Horizontal Gas Injection into Downward-Flowing Liquid, Metallurgical and Materials Transactions, Vol. 32B, pp. 1143-1159, Dec, 2001

Bauermeister M., Laue H., On the State of LMFBR Structural Analysis and Existing Rules with Regard to Class 1 Components of SNR 2, Nuclear Engineering and Design, Vol. 113, pp. 403-421, 1989

Becker A.A., Understanding Non-Linear Finite Elements Analysis Through Illustrative Benchmarks, NAFEMS, Glasgow, 2001

Bianchi F., Artioli C., Burn K.W., et al., Status and Trend of Core Design Activities for Heavy Metal Cooled Accelerator Driven System, Proceedings of ICENES'05, ISBN 90-7697-110-2, Aug, 2005

Buongiorno J. et al., Thermal Design of a Lead-Bismuth Cooled Fast Reactor with In-Vessel Direct Contact Steam Generator, Proceedings of ICONE'9, Nice, France, Apr, 2001

Carlsson J., Wider H., Emergency Decay Heat Removal by Reactor Vessel Auxiliary Cooling System from an Accelerator Driven System, Nuclear Technology, Vol. 140, No. 1, pp. 28-40, 2002

Ceballos C., Lathouwers D., Verkooijen A., Performance of Liquid Metals in Natural Circulation, Proceedings of ICAPP'04, Pittsburgh, USA, Jun, 2004

Ceballos C., Marcel C., Lathouwers D., et al., Experimental and Analytical Study of Buoyancy Enhancement with Two-Phase Gas-Liquid Metal in a Closed Loop, Proceedings of ICENES'05, ISBN 90-7697-110-2, Aug, 2005

Chang J.E., Suh K.Y., Hwang I.S., Natural Circulation Capability of Pb-Bi Cooled Fast Reactor: Peacer, Progress in Nuclear Energy, Vol. 37, No. 1-4, pp. 211-216, 2000

Chen X., Suzuki T., Rineiski A., et al., Source and Reactivity Perturbations in Accelerator Driven Systems with Conventional MOX and Advanced Fertile Free Fuels, Proceedings of Physor'04, Chicago, USA, Apr, 2004

Cheng S.K., Todreas N.E., Hydrodynamic Models and Correlations for Bare and Wire-Wrapped Hexagonal Rod Bundles-Bundle Friction Factors, Subchannel Friction Factors and Mixing Parameters, Nuclear Engineering and Design, Vol. 92, pp. 227-251, 1986

Cheng X, Cahalan J.E., Fink P.J, Safety Analysis of an Accelerator Driven Test Facility, Nuclear Engineering and Design, Vol 229, pp. 289-306, 2004

Choi S.K., Choi I.K., Nam H.Y., et al., Measurement of Pressure Drop in a Full-Scale Fuel Assembly of a Liquid Metal Reactor, Journal of Pressure Vessel Technology, Vol 125, pp. 233-238, 2003

Cinnoti L., Engineering Solutions and Thermalhydraulic Issues for a Promissing Safe ADS System, Proceedings of NUTHOS-6, Nara, Japan, Oct, 2004

Clift R., Grace J.R., Weber M.E., Bubbles, Drops and Particles, ISBN-0-12-176950-X, New York, Academic Press, 1978

Coddington P., Mikityuk K., Schikorr M., et al., Safety analysis of the EU-PDS-XADS Designs, NEA workshop on HPPA, Daejeon, Korea, May, 2004

Collier J.G., Thome J.R., Convective Boiling and Condensation, ISBN-0-19-856296-9, 3rd Edition, Oxford University Press, New York, 1999

Cometto M., Wydler P., Chawla R., A Comparative Physics Study of Alternative Long-Term Strategies for Closure of the Nuclear Fuel Cycle, Annals of Nuclear Energy, Vol 31, pp. 413-429, 2004

D'Angelo A., Gabrielli F., Benchmark of Beam Interruptions in an Accelerator Driven System, NEA/NSC/DOC17, Issy-les-Moulineaux, France, 2003

Davis C., Kim D., Todreas N., et al., Core Power Limits for a Lead Bismuth Natural Circulation Actinide Burner Reactor, Nuclear Engineering and Design, Vol 213, pp. 165-182, 2002

Duderstadt J., Hamilton L.J., Nuclear Reactor Analysis, Ed. John Wiley and Sons Inc., New York, USA, 1976

ENEA, A European Road Map for Developing Accelerator Driven Systems (ADS) for Nuclear Waste Incineration, ISBN 88-8286-008-6, Roma, 2001

Eriksson M., Reliability Assessment of The LANSCE Accelerator System, Dep. Nuclear & Reactor Physics - MSc. Thesis, Royal Institute of Technology, Stockholm, Sweden, 1998

Eriksson M., Cahalan J.E., Inherent shutdown Capabilities in Accelerator Driven Systems, Annals of Nuclear Energy, Vol 29, pp. 1689-1706, 2002

Eriksson M., Cahalan J.E., Yang W.S., On the Performance of Point Kinetics for the Analysis of Accelerator Driven Systems, Nuclear Science and Engineering, Vol 149, pp. 298 - 311, 2005

Eriksson M., Wallenius J., Jolkkonen M., et al., Inherent Safety of Fuels for Accelerator Driven Systems, Nuclear Technology, Vol 151, Sep, 2005
Femlab User Guide 3.0, Comsol Multiphysics, Dec, 2003

Ferziger J.H., Peric M., Computational Methods for Fluid Dynamics, Ed. Springer, Berlin, Germany, 2002

Fluent Inc, Fluent 6.1 UDF Manual, Lebanon, UK, 2003

Fomichenko P., Lead-cooled Reactors, New Concepts and Applications,Lecture Notes from the 2004 Frederic Joliot & Otto Hahn Summer School, Cadarache, France, Aug, 2004

Friedel L., Improved Friction Pressure Drop Correlations for Horizontal and Vertical Two-Phase Pipe Flow, European Two-Phase Flow Meeting, Paper E2, Ispra, Italy,Jun, 1979

Garland W.J., Butler M., Saunders F., Single-Phase Friction Factors for MNR Thermalhydraulic Modelling, Technical Report 1998-02, MacMaster University, Hamilton, Canada, Feb, 1999

Gudowski W., Arzhanov V., Broeders C., et al., Review of the European Project - Impact of Accelerator Based Technologies on Nuclear Fission Safety (IABAT), Progress in Nuclear Energy, Vol. 38, No. 1-2, pp. 135-151, 2001

Guet S., Bubble Size Effect on Gas-Lift Technique, ISBN-90-407-2492-X, Delft University Press, Delft, The Netherlands, 2004

Hetrick D.L., Dynamics of Nuclear Reactors, American Nuclear Society, Illinois, USA, 1993

Hoffman E.A., Stacey W.M., Comparative Fuel Cycle Analysis of Critical and Subcritical Fast Reactor Transmutation Systems, Nuclear Technology, Vol. 144, Oct, 2003

IAEA, The Safety of Nuclear Installations, Safety Series No. 10, Vienna, 1993

IAEA, Defence in Depth in Nuclear Safety, INSAG-10, Vienna, 1996

IAEA-TECDOC-1060, LMFR Core and Heat Exchanger Thermohydraulic Design: Former USSR and Present Russian Approaches, Vienna, Jan, 1999

IAEA-TECDOC-1289, Comparative Assessment of Thermophysical and Thermohydraulic Characteristics of Lead, Lead-Bismuth and Sodium Coolants for Fast Reactors, Vienna, Jun, 2002

Ilincev G., Research Results on the Corrosion Effects of Liquid Heavy Metals Pb, Bi and PbBi on Structural Materials with and without Corrosion Inhibitors, Nuclear Engineering and Design, Vol 217, pp. 167-177, 2002

Ishii M., Chawla T.C., Local Drag Laws in Dispersed Two-Phase Flow, ANL-79-105, Argonne National Laboratory, 1979

Jovanovich M., Dynamic Behavior and Stability Analysis of an Accelerator Driven System (ADS), Faculty of Mechanical Engineering - MSc. Thesis, Delft University of Technology, Delft, The Netherlands, 2004

Kataoka I., Ishii M., Drift Flux Model for Large Diameter Pipe and New Correlation for Pool Void Fraction, International Journal of Heat and Mass Transfer, Vol. 30, pp. 1927-1939, 1987

Kerdraon D., Billebaud A., Brissot R., et al., Characterization and Extrapolation of a Conceptual Experimental Accelerator Driven System, Progress in Nuclear Energy, Vol. 42, No. 1, pp. 11-24, 2003

Klueh R.L., Hashimoto N., et al., A Potential New Ferritic/Martensitic Steel for Fusion Applications, Journal of Nuclear Materials, Vol. 697, pp. 283-287, 2000

Lu Y.L., Chen L.J., Wang G.Y., et al., Hold time Effects on Low Cycle Fatigue Behavior of Haynes 230 Super-alloy at High Temperatures, Materials Science and Engineering, Vol. 409, pp. 282-291, 2005

Ludwig P.H., Verkooijen A., Thermal Stress in a Core of a Fast Energy Amplifier Cooled with Lead Under Normal Operating Conditions Including Reactor Scram and Secondary Pump Trip, Proceedings of the ICENES'00, Petten, The Netherlands, 2000

MacDonald P., Buongiorno J., Design of an actinide burning, lead or lead-bismuth cooled reactor that produces low cost electricity, INEEL/EXT-01-013376, MIT-ANP-PR-083, Oct, 2001

Mannan S.L., Valsan M., High -Temperature Low Cycle Fatigue, Creep-Fatigue and Thermomechanical Fatigue of Steels and their welds, International Journal of Mechanical Sciences, Vol. 48, pp. 160-175, 2005

Maschek W., Rineiski A., Flad M., et al., Analysis of Severe Accident Scenarios and Proposal for Safety Improvements for the ADS Transmuters with Dedicated Fuel, Nuclear Technology, Vol 141, Feb, 2003

Maschek W., Rineiski A., Suzuki T., et al., Safety aspects of oxide fuels for transmutation and utilization in accelerator driven systems, Journal of Nuclear Materials, Vol 320, pp. 147-155, 2003

Morita K., Maschek W., Flad M., et al., Thermophysical Properties of Lead-Bismuth Eutectic Alloy for Use in Reactor Safety Analysis, Second Meeting of the NEA Nuclear Science Committee, Working Group on LBE Technology, France, Sep, 2004

Nifenecker h., Meplan O., David S., Accelerator Driven Subcritical Reactors, ISBN-0-7503-0743-9, IOP Publishing, 2003

Nishi Y., Kinoshita I., Nishimura S., Experimental Study on Gas Lift Pump Performance in Lead-Bismuth Eutectic, Proceedings of ICAPP'03, Cordoba, Spain, May, 2003

Nishi Y., Kinoshita I., Nishimura S., Flow Characteristics of Lead-Bismuth Eutectic with Gas-Lift Pump, Proceedings of 6th ASME-JSME Thermal Engineering Joint Conference, TED-AJ03-536, Mar, 2003

OECD/NEA, Accelerator Driven Systems (ADS) and Fast Reactors (FR) in Advanced Nuclear Fuel Cycles - A Comparative Study, Paris, France, 2002

Ott K.O., Neuhold R.J., Introductory Nuclear Reactor Dynamics, ISBN 0-89448-029-4, American Nuclear Society, Illinois, USA, 1985

Patankar S.V., Numerical Heat Transfer and Fluid Flow, Hemisphere Publishing Corporation, New York, USA, 1980

Perdu F., Loiseaux J.M., Billebaud A., et al., Prompt Reactivity Determination in a Subcritical Assembly Through the Response to a Dirac Pulse, Progress in Nuclear Energy, Vol. 42, No. 1, pp. 107-120, 2003

Ponomarev L.I., High Power Accelerators in Fundamental Physics and Nuclear Technology, Progress in Nuclear Energy, Vol. 40, No. 3-4, pp. 655-660, 2002

Rineiski A., Maschek W., Kinetics Models for Safety Studies of Accelerator Driven Systems, Annals of Nuclear Energy, Vol 32, pp. 1348 - 1365, 2005

Rubbia C., Rubio J.A., Buono S., et al., Conceptual design of a fast neutron operated high power energy amplifier, CERN/AT/95-44(ET), Geneve, 1995

Rust J.H., Nuclear Power Plant Engineering, Haralson Publishing Company, Atlanta, USA, 1979

Rydin R.A., Woosley M.L., Evidence of Source Dominance in the Dynamic Behavior of Accelerator Driven Systems, Nuclear Science and Engineering, Vol 126, pp. 341-344, 1997

Salvatores M., Nuclear Fuel Cycle Strategies Including Partioning and Transmutation, Nuclear Engineering and Design, Vol. 235, pp. 805-816, 2005

Sasa T., Oigawa H., Tsujimoto K., et al., Research and Development on Accelerator Driven Transmutation System at JAERI, Nuclear Engineering and Design, Vol. 230, pp. 209-222, 2004

Satyamurthy P., Dixit N.S., Thiyagarajan T.K., et al., Two-Fluid Model Studies for High Density Two-Phase Liquid Metal Vertical Flows, International Journal of Multiphase Flow, Vol. 24, pp. 721-737, 1998

Sauzay M., Mottot M., Allais L., et al., Creep-Fatigue Behavior of an AISI Stainless Steel at 550 C, Nuclear Engineering and Design, Vol 232, pp. 219-236, 2004

Schikorr W.M., Assessments of the Kinetics and Dynamic Transient Behavior of Subcritical Systems (ADS) in comparison to Critical Reactor Systems, Nuclear Engineering and Design, Vol. 210, pp. 95-123, 2001

Seltborg P., Wallenius J., Tucek K., Gudowski W., Definition and Application of Proton Source Efficiency in Accelerator Driven Systems, Nuclear Science and Engineering, Vol. 145, pp. 390-399, 2003

Skrzypek J.J., Plasticity and Creep: theory, examples and problems, ISBN-0-8493-9936-X, CRC Press, Boca Raton, 1993

Spencer B., The rush to heavy liquid metal reactor coolants - gimmick or reasoned, Proceedings of ICONE'8, Baltimore, USA, Apr, 2000

Suzuki T., Tobita Y., Kondo S., et al., Analysis of Gas-Liquid Metal Two-Phase Flows Unsing a Reactor Safety Analysis Code SIMMER-III, Nuclear Engineering and Design, Vol. 220, pp. 207-223, 2003

Tak N., Neitzel H.J., Cheng X., Computational Fluid Dynamics Analysis of Spallation Target for Experimental Accelerator Driven Transmutation System, Nuclear Engineering and Design, Vol. 235, pp. 761-772, 2005

Takano H., Nishihara K., Symbiosis System for Transmutation on Nitride Fuel FBR and ADS, Progress in Nuclear Energy, Vol. 40, No. 3-4, pp. 473-480, 2002

Tang Y.S., Cofield Jr. R.D., Markley R.A., Thermal Analysis of Liquid Metal Fast Breeders, The American Nuclear Society, Illinois, USA, 1978

Timoshenko S., Goodier J.N., Theory of Elasticity, Ed. MacGraw-Hill, Tokyo, Japan, 1951

Todreas N.E., Kazimi M.S., Nuclear System I - Thermalhydraulic Fundamentals, Hemisphere Publishing Corporation, New York, USA, 1990

Uwaba T., Ukai S., Study on Short Term Stress Limit in Fast Reactor Fuel Pin Designs, Nuclear Engineering and Design, Vol 234, pp. 51-59, 2004

Van Dam H., Loss of Cooling Consequences in Critical and In Accelerator Driven Reactor Systems, Annals of Nuclear Energy, Vol. 26, pp. 1489-1495, 1999

Vickers L.D., Radiological Assessment of Target Materials for Accelerator Transmutation of Waste Applications, Nuclear Science and Engineering, Vol. 145, pp. 354-375, 2003

Waltar A.E., Reynolds A.B., Fast Breeder Reactors, Pergamon Press, New York, 1981

Welander P., On the Oscillatory Instability of a Differentially Heated Fluid Loop, Journal of Fluid Mechanics, Vol. 29, pp. 17-30, 1967

Yamamoto A., Shiroya S., Study on Neutronics Design of Accelerator Driven Subcritical Reactor as Future Neutron Source, part 1: Static Characteristics, Annals of Nuclear Energy, Vol 30, pp. 1409-1424, 2003

Zhukov A.V., Yu A. Kuzina, Smirnov V.P., et al., Heat Transfer and Temperature Distribution in Rod Bundles for LCFR, Proceedings Scientific Session MEPhI-2000, Vol. 8, pp. 108-110, 2000

Acknowledgment

In the first place, I would like to express my deepest gratitude to Prof. Adrian Verkooijen for giving me the opportunity to pursue this research and assisted me during this process. His guidance, interest, motivation and support in all different situations that arise during this 4 years are greatly appreciated.

Special thanks to Dr. Danny Lathouwers, who have got involved in this research voluntarily and become the daily supervisor of the project. His continuous assistance, support and dedication were important to develop further the project goals and the quality of reports. I would like to thank Prof. Tim van der Hagen for hosting me during 4 years inside his research group Physics of Nuclear Reactors. Within them, I have received a suitable working environment to study reactor safety. In similar way, I thank all the group of reactor physics for becoming my friends and colleagues during these 4 years.

Many thanks to the group of Energy Technology Department (nowadays – Process and Energy Department). They were my friends and colleagues during the M.Sc. and Ph.D. that I have followed in Delft. I do appreciate very much the shared moments and the respect we have for each other. Special thanks to the students Els Espeel and Marco Jovanovich for their contribution to my work and the beautiful experience of working together as student-supervisor relationship.

A very special acknowledgment to my friends Gianluca Di Nola, Christian Marcel, Hirokazu Shibata, Masahiro Furuya, Stavros Christoforou, Jose Maria Castillo, Martha Otero, Laura Rovetto, Hanneke Pieters, Marcela Velasquez and Ana Valencia. You all have been my family in this country; indeed, it would not have been possible without you. Also, many special thanks to more friends Elroy Powell, Rene Cox, Marthijn Kort, Alexander Bode, John Drissen, Camila Pinzon and Margarita Jans. With whom I have shared a lot of beautiful moments. I deeply value you.

Finally, I want to thank my family. My parents for the freedom, for the respect, for the love, for the support, for encouraging me to make sacrifices, to aim to be always a better person, to grow professionally, following always my dreams. My brother and sister and sister in law, for their interest, support and love.

Curriculum Vitae

The author Carlos Alberto Ceballos Castillo was born on July 3rd, 1974 in Bogota D.C., Colombia. He studied a Bachelor course in Mechanical Engineering at Universidad Nacional de Colombia (from 1991 to 1997), the academic year of 1996, he spent working at Occidental Petroleum Company. The year from July 1998 to 1999, he worked as a social volunteer helping disabled people in England. He returned to Colombia and worked for one year in Allianz-AGF as an engineer analyst and later he moved to the Netherlands to follow a M.Sc. course sponsored by Nuffic (from 2000 to 2002). His master thesis *"CFD modeling of co-firing secondary fuels with pulverized coal in a large scale boiler"* - developed with the support of E.On Benelux - enforce the significance for the industry and the Dutch government to research on co-combustion technology.
The Ph.D. work reported in this dissertation was accomplished between 2002-2006, when the author was appointed as a research assistant (AIO) in the group Physics of Nuclear Reactors, faculty of Applied Physics from Delft University of Technology. Currently, he is employed in the Central Engineering Department at NEM B.V. He develops tools for the design and dynamic analysis of power plants.